Lecture Notes
in Control and Information Sciences   260

Editors: M. Thoma · M. Morari

Springer-Verlag London Ltd.

Andreas Kugi

Non-linear Control Based on Physical Models

Electrical, Mechanical and Hydraulic Systems

With 47 Figures

 Springer

Series Advisory Board

Author

Andreas Kugi, Dr Techn, DipIng
Department of Automatic Control and Control Systems Technology,
Institute for Automatic Control and Electrical Drives,
Johannes Kepler University of Linz, Altenbergerstr. 69, 4040 Linz, Austria

ISBN 978-1-85233-329-4   ISBN 978-1-84628-554-7 (eBook)
DOI 10.1007/978-1-84628-554-7

British Library Cataloguing in Publication Data
Kugi, Andreas
    Non-linear control based on physical models : electrical,
    mechanical and hydraulic systems. - (Lecture notes in
    control and information sciences ; 260)
    1. Automatic control 2. Nonlinear control theory
    I. Title
    629.8'36
ISBN 978-1-85233-329-4

Library of Congress Cataloging-in-Publication Data
Kugi, Andreas, 1967-
    Non-linear control based on physical models : electrical, mechanical and hydraulic
    systems / Andreas Kugi.
        p. cm. – (Lecture notes in control and information sciences ; 260)
    Includes bibliographical references and index.
    ISBN 978-1-85233-329-4 (alk. paper)
        1. Automatic control. 2. Control theory. 3. Mechatronics. I. Title. II. Series.
TJ213 .K815 2000
629.8'36—dc21
                                                                      00-063761

Typesetting: Camera ready by author
69/3830-543210 Printed on acid-free paper SPIN 10765767

To

Erika and Marianne

Preface

Non-linear control is undoubtedly one of the most active research areas in the fields of automatic control and control systems technology. The research effort on non-linear control is getting more and more important because of the demanding performance required in practical applications and because most of the physical systems are non-linear in nature. Apart from the advances in non-linear control theory there is an increasing availability of computer programs for numeric and symbolic computation. Furthermore, the power of the automation hardware being used in the industry enables the real-time execution of the sometimes rather complicated non-linear control laws. Only the progress in all these areas makes the practical use of the non-linear control concepts possible.

Non-linear Control and Physical Models: All the electromagnetic, mechanical and hydraulic systems being considered in this work allow a strong analytic mathematical description, which is why, we focus our attention on so-called model-based non-linear control approaches. Now, the literature offers a variety of methods for the model-based non-linear controller design. The choice of the right control design strategy for a successful practical implementation depends strongly on the application, the demands on the closed loop, the restrictions and limitations of the plant itself and the actuators, the measurable quantities and its accuracy, the limitation of the real-time hard- and software platform etc. and cannot be answered generally and without preceding detailed investigation of the plant to be controlled. However, all the model-based non-linear control approaches have one fact in common, namely that somehow the knowledge of the underlying physical structure helps to solve the design problems. In order to stress this argument let us mention some examples, such as the solution of the Hamilton-Jacobi-Bellman-Isaacs equation/inequality for the non-linear H_∞-design; the solution of the Frobenius-type partial differential equations for the input-state or input-output exact linearization; the determination of the flat outputs within the flatness approach, or the suitable choice of the Lyapunov-like functions within the backstepping or composite Lyapunov design. This relation between non-linear control theory and physics is not new, it is rather rediscovered. Since the very beginning of the non-linear control theory certain physical observations have served as a starting point for a generalized mathematical theory, e.g., the direct method of Lyapunov, LaSalle's invariance principle, the notion of passivity and dissipativity etc. More recently, the so-called passivity-based control concepts went a step further. They do not only take advantage of certain physical properties, they even intend to design a non-linear controller in such a way that the physical structure is preserved in the closed loop.

Derivation of Physical Models and Non-linear Control: Not only the mathematical models themselves but also their step-by-step derivation, starting with the basic laws of physics, bring about a deep understanding of the physical background. The detailed setting up of the mathematical models enables a strict differentiation between the balance equations (e.g., energy or mass balance etc.), which, in general, hold exactly and the so-called constitutive equations (e.g., friction, leakage etc.), which contain normally many unknown parameters. It does not seem to be necessary to derive in a first step the mathematical model of a physical system and to try to explore its physical structure in a second step. This is why this work focuses on procedures where the mathematical model of a physical system is set up in a form which reveals directly the underlying physical structure as is the case for a port-controlled Hamiltonian system with and/or without dissipation, or PCH-/PCHD-system for short. Furthermore, it is absolutely necessary to point out clearly whenever a simplification of a mathematical model is performed or certain physical effects are neglected because such steps may destroy partially the physical structure.

Non-linear Control in the Industrial Environment: Despite all the advances in non-linear control, the number of realizations of non-linear controllers in the industrial environment is not as widespread as one might expect from the well established non-linear control theory. One possible explanation for this is that a straightforward application of the non-linear control methods often results in a closed loop system which is very sensitive to parameter variations and/or transducer and quantization noise. The control task is getting even more difficult since in general not all quantities are measurable. Furthermore, in contrast to linear systems, the separation property of an observer-controller based control design procedure is no longer valid in the non-linear case. Of course, also this work will not give a general solution to all these problems. But it will be shown for certain classes of applications, how a modified control approach, which takes into account the special features of the plant, can be successful. A prerequisite for a control concept to be practically feasible is that it is tested in advance on a simulator. The simulator must contain a much more detailed model than the model which serves as a basis for the controller design. In the simulator all the "dirty" effects, like the unmodelled dynamics of the sensors and actuators, the quantization, the transducer noise, the sampling process, stick-slip friction effects, leakage effects, parameter inaccuracies, aging induced changes of the system dynamics etc. have to be included. From our experience, a controller which can cope with all these "dirty" effects in the simulator, has a good chance to meet the requirements of the rough industrial environment.

Goal of the Work: The main purpose of this work is to elaborate the link between modelling and non-linear control, in particular for electromagnetic systems in Chapter 3, finite- and infinite-dimensional mechanical PCH-systems in Chapter 4 and hydraulic drive systems in Chapter 5. A great store is set by giving a unique mathematical formulation of the different disciplines involved, namely electrical, mechanical and hydraulic engineering. We always try to point out the common mathematical structure of the different physical models and this also makes it possible to make use of synergetic effects, like applying reliable control strategies from one discipline to the other. The work will also demonstrate, how the physics behind a mathematical model can contribute to the success of a certain control strategy. Furthermore, the practical relevance of the applications contained in this work in combination with a profound theory should protect from the reproach that non-linear control is solely of theoretical interest.

Organization of the Work: The first chapter is devoted to some basics of Lyapunov's theory, dissipativity, passivity, positive realness and absolute stability. Special emphasis is laid on elaborating the physical idea behind these concepts. In the last part of this chapter the basic structure of port-controlled Hamiltonian systems with and without dissipation, PCHD- and PCH-systems for short, will be discussed.

Chapter 2 summarizes the essential results of those non-linear model-based control approaches which will be used in the applications of this work and which have a more or less strong relation to the dissipativity and passivity concept. In particular, these are the non-linear state feedback H_2-design for affine-input systems, the non-linear state feedback H_∞-design for affine input systems and the passivity-based control concept. For a special type of affine-input systems the non-linear state feedback H_2-design is extended in such a way that an integral part can be systematically included in the controller.

In Chapter 3 a theory for an energy-based description of electric networks, which can be regarded as an extension of the well-known theory of Brayton-Moser, is formulated. A combination of this approach with graph theory allows us to set up the network equations directly in the form of a PCHD-system. The method being presented is applicable for non-linear two- and three-phase systems with and without dependent sets of inductor currents and/or capacitor voltages. The big advantage of this approach, in particular for three-phase systems, is that it does neither require a linear magnetic characteristic nor any assumption for the spatial distribution of the coupling inductors or capacitors. Furthermore, this technique is used for the calculation of the average model of PWM (pulse-width-modulation)-controlled electric circuits with bipolar switching, where the duty ratio is the control input. Depending on the location of the switch(es), different energy flows of the PWM-controlled system can be influenced by changing the duty ratio. Among other examples, it is shown by means of the laboratory model of a special dc-to-dc converter, namely the Čuk-converter, how the presented theory can contribute to the solution of the non-linear H_2-control design with and without integral term. In the last part of Chapter 3, the co-energy concept is introduced to calculate the magnetic and electric coupling forces of electromechanical systems.

Chapter 4 describes finite- and infinite-dimensional mechanical systems which have the structure of a PCH-system. In order to obtain a uniform description of the finite- and infinite-dimensional case, the mathematical models are founded on the Poisson-bracket form of the equations of motion. It turns out that the Hamiltonian structure offers some pleasing properties which can be advantageously used for the controller design. In particular, the non-linear H_2-, the non-linear H_∞-design, the PD-(proportional differential) controller design and the idea of disturbance compensation will be adapted for finite- and infinite-dimensional PCH-systems. The different control strategies developed so far for PCH-systems will be applied to an infinite-dimensional piezo-electric composite beam structure. The feasibility of these control concepts relies on the fact that the piezoelectric structures allow a spatial distribution of the piezoelectric sensor and actuator layers. The design of the spatial pattern of the sensor and actuator electrodes is an additional degree of freedom and can be regarded as a part of the control synthesis task. In this way, we are able to design the sensor and actuator layers such that they are collocated and hence the well known effects of observation/actuation spillover can be prevented.

In Chapter 5 two special types of hydraulic drive systems, namely a valve-controlled piston actuator and a pump-displacement controlled rotational piston actuator, are discussed. The underlying physical structure is again elaborated carefully and this knowledge is advantageously used for the controller design. The mathematical model being considered for the valve-controlled piston actuator has the pleasing property that it is exact input-state linearizable. But in some industrial applications, it turns out that those controllers, which have to rely on the knowledge of the piston velocity, have problems in the case of noisy measurements and/or parameter variations. This is also why we propose a non-linear controller based on the input-output linearization which requires only measurable quantities. The feasibility of this non-linear control concept will be demonstrated by means of the HGC (hydraulic gap control), which is the innermost control loop of the thickness control concept in rolling mills. Irregularities in the mill rolls and/or roll bearings may cause so-called roll eccentricities which appear as periodic disturbances in the strip exit thickness. In general, these disturbances cannot be eliminated by means of the conventional thickness control concepts, like the HGC. Therefore, an adaptive controller is developed to compensate periodic disturbances with a known period but an unknown phase and amplitude. A passivity based argument is used to prove the stability. The second part of this chapter is concerned with a closed-coupled hydrostatic drive unit consisting of a variable-displacement axial-piston pump and a fixed-displacement axial-piston motor. It will be shown that the mathematical model of the hydrostatic drive unit has the same mathematical structure as certain types of PWM-controlled dc-to-dc converters. A detailed mathematical model of the swash-plate mechanism of the variable-displacement pump is derived. The complexity of this model is gradually reduced on the basis of physical considerations. The so-obtained simpler model is used to design an on-line discrete open-loop observer for the swash-plate angle. Finally, comparative results of the measured and the estimated swash-plate angle for an industrial drive box are presented.

Acknowledgments: This work is a result of my research in the field of non-linear control at the Department for Automatic Control and Control Systems Technology within the Division of Mechatronics at the Johannes Kepler University of Linz, Austria and it also served as a basis for my Habilitation. Many colleagues, industrial partners, friends and students have contributed more or less to this work. In particular, I want to express my deepest gratitude to Professor Kurt Schlacher for his advice and help during all the years at the department. Since I started my scientific career as a young postgraduate student at his department, he not only shaped my scientific attitude but also gave me the necessary freedom for going my own scientific ways. I am also thankful to my (former) colleagues Reinhard Gahleitner, Werner Haas (STEYR Antriebstechnik GmbH & Co OHG, Steyr, Austria), Mathias Meusburger (VOEST-ALPINE Technology SAT, Vienna, Austria), Kurt Zehetleitner at the Department and Gernot Grabmair, Stefan Fuchshumer and Rainer Novak at the Christian Doppler Laboratory for the Automation of Mechatronic Systems in Steel Industries for many fruitful discussions. Rainer Novak generously helped me with the final preparation of the rolling mill application in Chapter 5. I am also indebted to Professor Hans Irschik from the Department of Technical Mechanics and Foundations of Machine Design, Johannes Kepler University of Linz, Austria, for the inspiring and fruitful cooperation that led to some essential results of Chapter 4. Here, I also would like to acknowledge my friend Manfred Kaltenbacher from the Department of Sensor Technology, University of Erlangen, Germany, for many years of interesting scientific discussions. Further, I wish to thank all professors, teaching staff and students of the Mechatronics Division for the good atmosphere, in particular, many thanks to all my students for their penetrating questions and the many years of wonderful discussions.

A rather strong influence on this work came from the experience of making non-linear control work in many industrial projects. In this context I would like to thank all our industrial partners for making the industrial problems more accessible to me. In particular, I would like to acknowledge Karl Aistleitner, Reinhold Fogel, Helge Frank, Georg Keintzel and Roman Schneeweiß from VOEST-ALPINE Industrieanlagenbau GmbH, Linz, Austria, and Heinz Aitzetmüller, Gottfried Hirmann and Heimo Nakesch from

STEYR Antriebstechnik GmbH & Co OHG, Steyr, Austria for their support to the applications of Chapter 5. I am also thankful to my friend Gernot Druml from A. Eberle GmbH, Nürnberg, Germany for stimulating, sometimes emotional discussions concerning the proper description of three-phase systems and for his influence on Chapter 3.

I am also indebted to Gill and Tom Gajdátsy, Darlington, UK, for their linguistic advice and for improving the textual quality. Many thanks to my parents for their love and their support all over the years - Liebe Eltern, ich danke Euch für Eure Liebe und Hilfe. Last but not least I am deeply grateful to my wife Erika for her never ending patience and love and for my little daughter Marianne.

The author would be grateful for reports of errors in this work to be sent to the following address:

Department for Automatic Control and Control Systems Technology
Johannes Kepler University of Linz
Altenbergerstr. 69, A-4040 Linz, Austria, Europe
email: kugi@mechatronik.uni-linz.ac.at

An up-to-date errata list will be published on the WWW-page
http://regpro.mechatronik.uni-linz.ac.at/springer_book/

Finally, I also thank Hannah Ransley and Nicholas Pinfield from Springer-Verlag London for their support and the pleasant cooperation.

Linz, July 2000   *Andreas Kugi*

Contents

1. Fundamentals

In this chapter, we briefly discuss some basics of Lyapunov's theory, dissipativity, passivity, positive realness and absolute stability. Thereby, special emphasis is laid on examining more closely the physical background and on elaborating the connections between these concepts. It should be mentioned that the intention of this chapter is not to present a complete theory; it rather summarizes those theoretical concepts which will be required in the subsequent chapters. In this sense some results are stated without proof, but throughout the whole chapter the reader is always referred to the corresponding literature for more details and information.

1.1 Stability of Equilibria

The stability of equilibrium points is essentially characterized in the sense of Lyapunov's theory. Consider a time-invariant autonomous system

$$\frac{\mathrm{d}}{\mathrm{d}t} x = f(x) \tag{1.1}$$

where $x \in R^n$ and f is a continuous vector field. Let us assume that (1.1) has a unique solution for any initial condition $x_0 = x(t_0)$ and for all $t \geq t_0$. This can be guaranteed if f satisfies a global Lipschitz condition [59], [144]. Without restriction of generality we assume that the origin $\bar{x} = 0$ is the equilibrium of (1.1), i.e. $f(\bar{x}) = 0$, whose stability has to be investigated. In the case $\bar{x} \neq 0$ we can always perform a suitable change of coordinates $z = x - \bar{x}$.

Definition 1.1. *Let $\varphi_t^f(x)$ denote the flow of (1.1) then the equilibrium is stable in the sense of Lyapunov if and only if, for each $\varepsilon > 0$, there is a $\delta(\varepsilon) > 0$ such that*

$$\|x_0\| < \delta \Rightarrow \left\|\varphi_t^f(x_0)\right\| < \varepsilon \tag{1.2}$$

for all $t \geq t_0$. If in addition a $\zeta > 0$ can be found such that

$$\|x_0\| < \zeta \Rightarrow \lim_{t \to \infty} \varphi_t^f(x_0) = 0 \tag{1.3}$$

holds then the equilibrium is said to be asymptotically stable in the sense of Lyapunov.

Here it is worth mentioning that the norm $\| \cdot \|$ in (1.2) and (1.3) is arbitrary since all norms on R^n are topologically equivalent. It is quite clear that the above stability definitions are not suitable for determining the stability character of an equilibrium. But this is precisely the genius of Lyapunov's theory: that it offers a possibility to investigate the stability without calculating the trajectories of the system (1.1). This leads us to the well known direct or second method of Lyapunov. Before giving a formulation of Lyapunov's direct method we will look at the underlying physical idea.

1.1.1 Physical Observation I

The basic idea of Lyapunov's direct method comes from the physical observation that the total stored energy of a mechanical or electrical system without external inputs is non-increasing in time. Furthermore, if the system contains dissipative elements, the total stored energy is monotonically decreased and hence even goes to zero. Let us consider the very simple electric circuit of Fig. 1.1 with the inductor L, the capacitor C and the resistances R_1 and R_2. The network equations read as

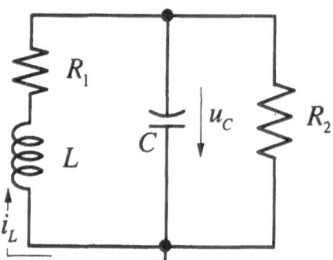

Fig. 1.1. Simple electric circuit.

$$
\begin{aligned}
\frac{\mathrm{d}}{\mathrm{d}t}i_L &= \frac{1}{L}\left(-u_C - R_1 i_L\right) \\
\frac{\mathrm{d}}{\mathrm{d}t}u_C &= \frac{1}{C}\left(i_L - \frac{u_C}{R_2}\right)
\end{aligned}
\tag{1.4}
$$

with the inductor current i_L and the capacitor voltage u_C as the state variables $x^T = [i_L, u_C]$. Obviously, $\bar{x} = 0$ is the only equilibrium of (1.4). The positive definite energy stored in the inductor and capacitor is given by

$$
V = \frac{1}{2}L i_L^2 + \frac{1}{2}C u_C^2
\tag{1.5}
$$

and the change of the stored energy V due to the motion of the system (1.4) takes the form

$$\frac{\mathrm{d}}{\mathrm{d}t}V = -R_1 i_L^2 - \frac{1}{R_2} u_C^2 \tag{1.6}$$

and is, in fact, negative definite. Thus, we may deduce from (1.5) and (1.6) that the equilibrium $\bar{x} = 0$ is asymptotically stable in the sense of Definition 1.1.

1.1.2 Mathematical Formulation I: Lyapunov's Direct Method

A generalization of this approach to the class of time-invariant autonomous systems (1.1) is known as the direct or second method of Lyapunov.

Theorem 1.1. *The equilibrium $\bar{x} = 0$ of (1.1) is stable in the sense of Lyapunov if there exists a continuously differentiable positive definite function $V(x)$ on a neighborhood $\mathcal{X} \subset R^n$ of 0 such that the relation*

$$\frac{\mathrm{d}}{\mathrm{d}t}V(x) = L_f V(x) \leq 0 \tag{1.7}$$

holds for all $x \in \mathcal{X}$ with $L_f V$ as the Lie derivative of V along the vector field f. Moreover, if

$$\frac{\mathrm{d}}{\mathrm{d}t}V(x) = L_f V(x) < 0 \tag{1.8}$$

for all $x \in \mathcal{X} - \{0\}$ the equilibrium is asymptotically stable. The function $V(x)$ is commonly referred to as a Lyapunov function.

In this sense the Lyapunov function can be regarded as an extension of the total stored energy of physical systems to a more general class. Therefore, it is obvious that within the controller design for physical systems the total stored energy serves as an appropriate Lyapunov function candidate. This is also why an energy based formulation of physical systems, like the Euler-Lagrange formulation of electromechanical systems, is so popular within the control community. We will emphasize this aspect in Chapters 3, 4 and 5 for electrical, electromechanical, mechanical and hydraulic systems.

1.1.3 Physical Observation II

In many physical applications, however, it turns out that the equilibrium $\bar{x}$ is asymptotically stable although; the time derivative of $V(x)$ in Theorem 1.1 is only negative semi-definite. As an example we will investigate the simple spring-mass-damper system of Fig. 1.2. Suppose the spring has the non-linear restoring force $F_c = \psi_F(z)$ where the function $\psi_F(z)$ satisfies the sector

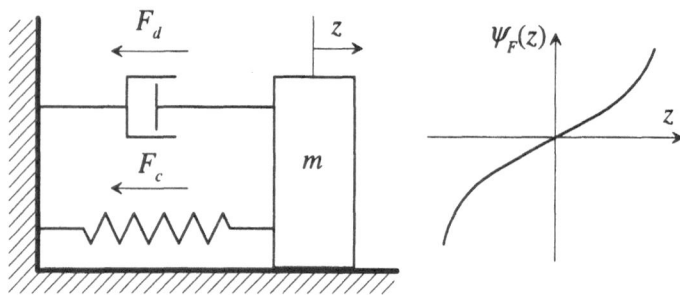

Fig. 1.2. Simple mechanical system.

condition $k_1 z^2 \le \psi_F(z)\, z \le k_2 z^2$ with $0 < k_1 < k_2$. The damping force is assumed to be proportional to the velocity, $F_d = d\frac{\mathrm{d}}{\mathrm{d}t}z(t)$, with the damping coefficient $d > 0$. Then the equations of motion are given by

$$\frac{\mathrm{d}}{\mathrm{d}t}z = v$$
$$\frac{\mathrm{d}}{\mathrm{d}t}v = -\frac{1}{m}\left(\psi_F(z) + dv\right)$$

(1.9)

with the state $x^T = [z, v]$ and $\bar{x} = 0$ as the only equilibrium. The kinetic and potential energy stored in the system

$$V = \frac{1}{2}mv^2 + \int_0^z \psi_F(w)\,\mathrm{d}w$$

(1.10)

is again a suitable Lyapunov function candidate. Taking the time derivative of V along an integral curve of (1.9), we see that

$$\frac{\mathrm{d}}{\mathrm{d}t}V = -dv^2$$

(1.11)

is only negative semi-definite. Following Theorem 1.1, we may deduce that the origin is stable but not asymptotically stable. But the stored energy V is decreasing everywhere except when $\frac{\mathrm{d}}{\mathrm{d}t}V = 0$, that is for $v = 0$ and z arbitrary. But from (1.9) one can easily conclude that the trajectory cannot be confined to the set of points described by $\frac{\mathrm{d}}{\mathrm{d}t}V = 0$, unless $z = 0$. In other words, the energy stored in the system will be dissipated until it is zero. This allows the conclusion that the origin is asymptotically stable. The exact mathematical explanation of this physical observation is given by the famous invariance principle of LaSalle.

1.1.4 Mathematical Formulation II: LaSalle's Invariance Principle

The invariance principle of LaSalle uses the concept of invariant sets. A set $\mathcal{V}$ is called a positively (negatively) invariant set of (1.1) if for each initial condition $x_0 = x(t_0) \in \mathcal{V} \Rightarrow \varphi_t^f(x_0) \in \mathcal{V}$ for all $t \ge t_0$ ($t \le t_0$).

Theorem 1.2. *Let $\mathcal{X} \subset R^n$ be a compact, positively invariant set of (1.1) and suppose $V(x) : \mathcal{X} \to R$ is a continuously differentiable function with $\frac{\mathrm{d}}{\mathrm{dt}} V(x) \le 0$ in $\mathcal{X}$. Further, let $\mathcal{V}$ be the largest positively invariant set of $W = \{x \in \mathcal{X} | \frac{\mathrm{d}}{\mathrm{dt}} V(x) = 0\}$ then, every solution of (1.1) starting in $\mathcal{X}$ approaches $\mathcal{V}$ as $t \to \infty$.*

Since the level set $\mathcal{X}_c = \{x \in \mathcal{X} | V(x) \le c\}$ of a positive definite function $V(x)$ with the property $\frac{\mathrm{d}}{\mathrm{dt}} V(x) \le 0$ is positively invariant and compact for a sufficiently small positive constant c, Theorem 1.2 can be extended in accordance with Theorem 1.1 in the form of the following corollary.

Corollary 1.1. *The equilibrium $\bar{x} = 0$ of (1.1) is asymptotically stable in the sense of Lyapunov if there exists a continuously differentiable positive definite function $V(x)$ on a neighborhood $\mathcal{X} \subset R^n$ of 0 such that the relation*

$$\frac{\mathrm{d}}{\mathrm{dt}} V(x) = L_f V(x) \le 0 \qquad (1.12)$$

holds for all $x \in \mathcal{X}$ and the largest positively invariant set $\mathcal{V}$ of $W = \{x \in \mathcal{X} | \frac{\mathrm{d}}{\mathrm{dt}} V(x) = 0\}$ contains only the equilibrium 0, i.e., $\mathcal{V} = \{0\}$.

The next remarks briefly deal with the global stability property, Lyapunov's theory for time-varying systems and with Lyapunov's theory for the infinite-dimensional case.

Remark 1.1. The stability concepts introduced so far are of a local nature only. In order to get global stability properties the function $V(x)$ in Theorem 1.1 and Corollary 1.1 must satisfy an extra condition, namely $V(x)$ must be radially unbounded. This means that $V(x) \to \infty$ as $\|x\| \to \infty$.

Remark 1.2. The basic results of Lyapunov's theory can also be extended to time-varying systems of the form

$$\frac{\mathrm{d}}{\mathrm{dt}} x = f(t, x) \qquad (1.13)$$

but this is not intended within this work. The reader is asked to refer to, e.g., [59], [135], [144].

Remark 1.3. In order to use Lyapunov's theory for infinite-dimensional systems, some new aspects have to be taken into account and its application may involve some rather delicate mathematical technicalities. The main reason for this is that in contrast to finite-dimensional systems in the infinite-dimensional case, the compactness of the level sets of the Lyapunov function is no longer automatically ensured. This property must be checked separately. For more information on this topic it is advisable to consult, e.g., [1], [82], [92].

1.1.5 Exponential Stability

Apart from stability in the sense of Lyapunov we will also need the notion of exponential stability, but now formulated for the non-linear time-varying case (1.13). Again, without loss of generality, let the origin $\bar{x} = 0$ be an equilibrium of (1.13), i.e. $f(t,0) = 0$ for all $t \geq 0$. Then the equilibrium is said to be exponentially stable if there exist positive constants α_1 and α_2 such that

$$\left\| \varphi_{t,t_0}^f(x_0) \right\| \leq \alpha_1 \|x_0\| \exp(-\alpha_2 t) \tag{1.14}$$

holds for all times $t \geq t_0$ and all initial conditions $x_0 = x(t_0) \in \mathcal{X} = \{ x \in R^n | \|x\| < r \}$ where $t_0 \geq 0$ and r is a suitable positive constant. Further, the equilibrium is globally exponentially stable if $\mathcal{X} = R^n$.

Theorem 1.3. *The equilibrium $\bar{x} = 0$ of (1.13) is exponentially stable if there exist a continuously differentiable function $V(t,x) : R_+ \times \mathcal{X} \to R$ and positive constants β_1, β_2, β_3 and $p \geq 1$ such that the relations*

$$\beta_1 \|x\|^p \leq V(t,x) \leq \beta_2 \|x\|^p$$
$$\frac{d}{dt} V(t,x) \leq -\beta_3 \|x\|^p \tag{1.15}$$

are satisfied for all $t \geq 0$ in a neighborhood $\mathcal{X}$ of 0. Moreover, if $\mathcal{X} = R^n$ the equilibrium is even globally exponentially stable.

For a proof, see e.g., [59], [144].

1.2 Dissipativity, Passivity and Positive Realness

1.2.1 Physical Observation III

From the first law of thermodynamics it is known that energy is neither "created" nor "destroyed", but only changes its form and all energy changes must cancel each other out at any instant in time. Next we will apply this very general energy balance principle to two specific cases, namely to a class of heat transfer systems and to electromechanical systems.

Heat Transfer Systems. At first let us consider a heat transfer system where the electromechanical effects may be neglected and hence the changes in the energy storage are solely due to changes in the internal thermal energy. From the energy conservation requirement the change in the thermal energy storage V follows the relation

$$\frac{d}{dt} V = p_{in} - p_{out} \tag{1.16}$$

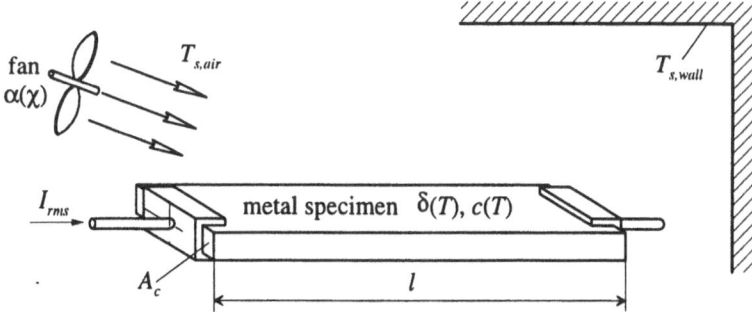

Fig. 1.3. Annealing of metal specimen.

with p_{in} and p_{out} as the energy flows which enter and leave the system.

As a typical application example we will investigate the annealing of a metal sheet specimen by means of conductive heating and forced convection as presented in Fig. 1.3 (see also [63] for details). By passing an electric current with the rms-value I_{rms} through the specimen thermal energy is generated due to Ohmic heating. Either a fan or compressed air is used to provide forced convection air cooling. We assume that at any time t the temperature T in the metal sheet specimen is uniform and that the surface of the surrounding walls is large in comparison to the specimen surface. Further, we will ignore the effect of heat conduction.

The internal thermal energy V of the specimen is given by

$$V(T) = c(T) mT \tag{1.17}$$

with the constant mass m and the specific heat $c(T)$ as a strictly increasing function of the temperature T. Following Ohm's law, we get the energy flow due to electric resistance heating in the form

$$p_{in} = I_{rms}^2 \delta(T) \frac{l}{A_c} \tag{1.18}$$

with the specific resistance $\delta(T)$, the length of the specimen l and the cross section A_c. The energy flows caused by forced convection and radiation are given by

$$p_{out,1} = \alpha(\chi) A_s (T - T_{s,air}) \tag{1.19}$$

and

$$p_{out,2} = \varepsilon \sigma A_s (T^4 - T_{s,wall}^4) \tag{1.20}$$

with the surface area A_s of the specimen, the temperature of the surrounding air and surrounding walls, $T_{s,air}$ and $T_{s,wall}$, respectively, the emissivity ε, the

Stefan-Boltzmann constant $\sigma = 5,67 \cdot 10^{-8}$ Wm^{-2}K^{-4} and the convection heat transfer coefficient $\alpha(\chi)$. Thereby, χ stands for the angular velocity of the fan or the pressure of the compressed air depending on the cooling equipment used. If we only have free convection the heat transfer coefficient $\alpha(\chi)$ is constant and lies within the range $2 - 25$ Wm^{-2}K^{-1} (see, e.g., [48]).

The dynamic system equation can be directly obtained by inserting (1.17)-(1.20) in (1.16) with the temperature T of the specimen as the state and the input $u^T = [I_{rms}, \chi, T_{s,air}, T_{s,wall}]$. Thus for a given input u the energy balance of (1.16) integrated from time t_0 to t along the integral curve of the system with the initial temperature $T(t_0)$ results in

$$V(T(t)) - V(T(t_0)) = \int_{t_0}^{t} s(I_{rms}, \chi, T_{s,air}, T_{s,wall}, T) \, d\tau \qquad (1.21)$$

with

$$s(I_{rms}, \chi, T_{s,air}, T_{s,wall}, T) =$$
$$I_{rms}^2 \delta(T) \frac{l}{A_c} - \alpha(\chi) A_s (T - T_{s,air}) - \varepsilon \sigma A_s \left(T^4 - T_{s,wall}^4\right) . \qquad (1.22)$$

Equation (1.21) says that the thermal energy stored in the specimen V at time t equals the stored energy at time t_0 plus the energy supplied or taken from the specimen with the so-called supply rate $s(I_{rms}, \chi, T_{s,air}, T_{s,wall}, T)$.

Electromechanical Systems. In the second case we consider electromechanical systems where the changes in the internal thermal energy are neglected. Here the rate of change of the energy V stored in the electromechanical system can be written in the form

$$\frac{d}{dt} V = p_{in} - p_{out} - p_{diss} \qquad (1.23)$$

where p_{in} and p_{out} denote the energy flows which enter and leave the system across its boundaries and p_{diss} is the energy flow which is dissipated into heat. As an example let us consider the electromagnetic valve of Fig. 1.4 which consists of a cylindrical core and a cylindrical plunger with mass m and diameter D moving in a guide ring. The coil, which has N turns with the total internal resistance R, is supplied by the voltage source U_0. Let us assume that the material of the core and plunger is infinitely permeable whereas the permeability of the guide ring is the same as that of air. Further, we assume that $h \ll D$ and $\delta \ll h$ and hence we will neglect all leakage effects. Then the energy stored in the magnetic circuit $\hat{w}_L(z, i_L)$ as a function of the plunger position z and the coil current i_L can be calculated in the form

$$\hat{w}_L = \frac{1}{2} L(z) i_L^2 \qquad (1.24)$$

with the inductance of the magnetic circuit

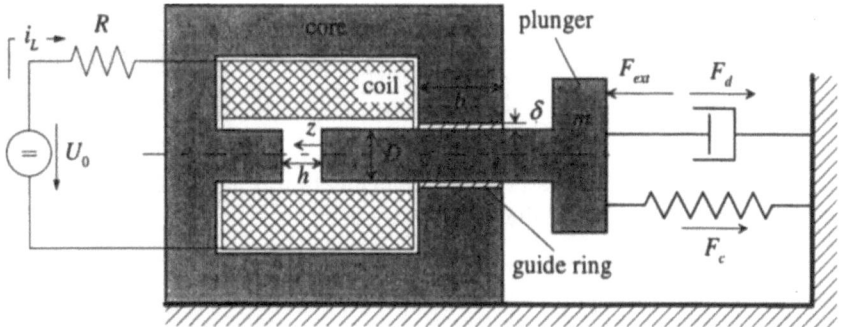

Fig. 1.4. Schematic diagram of an electromagnetic valve.

$$L(z) = \frac{\mu_0 N^2 D^2 \pi \,(D+\delta)\,\pi b}{4\,(h-z)\,(D+\delta)\,\pi b + \delta D^2 \pi} \tag{1.25}$$

and $\mu_0 = 4\pi \cdot 10^{-7}\ \text{VsA}^{-1}\text{m}^{-1}$ as the permeability of air. Since in this example the energy $\hat{w}_L$ and the co-energy $\check{w}_L$ of the magnetic circuit are equal the magnetic force is given by the relation

$$F_{mag} = \frac{\partial}{\partial z}\check{w}_L = \frac{1}{2}\frac{\partial L(z)}{\partial z}i_L^2 \ . \tag{1.26}$$

A detailed treatment of the energy/co-energy concept for electromechanical systems is presented in Section 3.6. The friction between the plunger and the guide ring is assumed to be proportional to the plunger velocity and it is supposed to be contained in the damping coefficient $d > 0$ of the damping force $F_d = d\frac{\mathrm{d}}{\mathrm{d}t}z\,(t)$. In addition to the magnetic force and the damping force, an external load force F_{ext} and a restoring force F_c due to a spring with the spring constant $c > 0$ act on the plunger. Hence, the system equations are of the form

$$\begin{aligned}
\frac{\mathrm{d}}{\mathrm{d}t}z &= v \\
\frac{\mathrm{d}}{\mathrm{d}t}v &= \frac{1}{m}\left(\frac{1}{2}\frac{\partial L(z)}{\partial z}i_L^2 - cz - dv + F_{ext}\right) \\
\frac{\mathrm{d}}{\mathrm{d}t}i_L &= \frac{1}{L(z)}\left(U_0 - Ri_L - \frac{\partial L(z)}{\partial z}i_L v\right)
\end{aligned} \tag{1.27}$$

where $x^T = [z, v, i_L]$ is the state, $u^T = [U_0, F_{ext}]$ the input and $L(z)$ the inductance from (1.25). The total energy V stored in the system is the sum of the magnetic energy $\hat{w}_L$ of (1.24), the kinetic energy of the plunger mass and the potential energy of the spring

$$V = \frac{1}{2}\left(L(z)i_L^2 + mv^2 + cz^2\right) \ . \tag{1.28}$$

The change of the stored energy V due to the motion of the system reads as

$$\frac{\mathrm{d}}{\mathrm{d}t} V = \underbrace{U_0 i_L + F_{ext} v}_{p_{in} - p_{out}} - \underbrace{\left(d v^2 + R i_L^2 \right)}_{p_{diss}} . \tag{1.29}$$

Integrating (1.29) for given input u from time t_0 to t along the integral curve of (1.27) with the initial value $x(t_0)$, we get

$$V(x(t)) - V(x(t_0)) \leq \int_{t_0}^{t} s(U_0, F_{ext}, i_L, v) \, \mathrm{d}\tau \tag{1.30}$$

with the supply rate

$$s(U_0, F_{ext}, i_L, v) = U_0 i_L + F_{ext} v . \tag{1.31}$$

1.2.2 Mathematical Formulation III: The Notion of Dissipativity

This physical observation of energy balance can now be embedded in the more general mathematical concept of dissipativity (see, e.g., [126], [143], [145]). Consider a dynamic state-space system of the form

$$\begin{aligned} \frac{\mathrm{d}}{\mathrm{d}t} x &= f(x, u) \\ y &= h(x, u) \end{aligned} \tag{1.32}$$

with the state $x \in \mathcal{X} \subset R^n$, the control input $u \in \mathcal{U} \subset R^m$ and the output $y \in \mathcal{Y} \subset R^p$. We will denote all those input functions $u(t)$ that determine the state $x(t)$ unambiguously for any initial value $x(t_0) = x_0$ and all $t \geq t_0$ as admissible. Let us define a real-valued function $s(u, y) : \mathcal{U} \times \mathcal{Y} \to R$, the so-called supply rate, such that $\int_{t_0}^{t} |s(u, y)| \, \mathrm{d}\tau < \infty$ for any $x_0 \in \mathcal{X}$, any admissible input u and all $t \geq t_0$.

Definition 1.2. *The system (1.32) is said to be dissipative with supply rate s if there exists a non-negative function $V(x) : \mathcal{X} \to R$ such that the inequality*

$$V(x(t)) - V(x(t_0)) \leq \int_{t_0}^{t} s(u(\tau), y(\tau)) \, \mathrm{d}\tau \tag{1.33}$$

holds for all admissible u, all initial values $x_0 \in \mathcal{X}$ and all $t \geq t_0$. The function $V(x)$ is also called a storage function and the inequality (1.33) is often referred to as dissipation inequality. If (1.33) holds with equality, the system (1.32) is said to be lossless with respect to the supply rate s.

At this point it should be emphasized that here and henceforth with $x(t)$ and $y(t)$ we mean $x(t) = \varphi(t, x_0, u(t))$ and $y(t) = h(\varphi(t, x_0, u(t)), u(t))$ where $\varphi(t, x_0, u(t))$ is the solution of (1.32) corresponding to the initial condition $x(t_0) = x_0$ and to the input $u(t)$, evaluated at time t.

Remark 1.4. In the sense of Definition 1.2 the heat transfer system of Fig. 1.3 is lossless with respect to the supply rate (1.22) and the electromechanical system (1.27) of Fig. 1.4 is dissipative with supply rate (1.31).

Among all possible storage functions $V(x)$ we shall be interested in the so-called available storage $V_a(x)$, defined by

$$V_a(x) = \sup_{u \in \mathcal{U}, \, t \geq t_0} \left(-\int_{t_0}^t s(u(\tau), y(\tau)) \, d\tau \right) \quad \text{with} \quad x(t_0) = x . \quad (1.34)$$

In [145] (see also [17], [143]) it is shown that the system (1.32) is dissipative with supply rate s if and only if $V_a(x)$ is finite for all $x \in \mathcal{X}$. If this is the case, the available storage $V_a(x)$ can be interpreted as the lower bound of all possible storage functions $V(x)$, i.e.

$$0 \leq V_a(x) \leq V(x) \quad (1.35)$$

for all $x \in \mathcal{X}$. A physical interpretation of $V_a(x)$ is that it gives the maximum amount of energy that can be extracted from the system with the initial condition $x(t_0) = x$.

Remark 1.5. If the storage function $V(x)$ is continuously differentiable in x then we can calculate the change of $V(x)$ along the trajectories of the system (1.32). Thus we obtain the so-called differential dissipation inequality associated to (1.33) in the form

$$\frac{d}{dt} V(x) \leq s(u(t), y(t)) . \quad (1.36)$$

This representation is useful in many situations, especially those concerning stability investigations.

Remark 1.6. Dissipativity is strongly related to the input-output stability of dynamic systems. In particular, the system (1.32) has L_2-gain less equal $\gamma > 0$ if it is dissipative with supply rate $s(u, y) = \gamma \|u\|^2 - \|y\|^2$ where $\| \ \|$ is the Euclidean norm. For a comprehensive presentation of the connections between input-output stability and dissipativity the reader is referred to a very recent book [143].

1.2.3 Passivity

Originally, the notion of passivity comes from the theory of electrical networks. In this respect, an electrical network is said to be passive if and only if it is composed entirely of positive resistors, inductors and capacitors (see, e.g., [19]). However, within the area of control and systems theory a passivity formulation for a much more general class of dynamic systems was developed. Subsequently, we will recapitulate some of the fundamental results of the passivity concept, which will be of importance for the next chapters. For this purpose, let us consider a dynamic system (1.32) with the number of inputs m being equal to the number of outputs p. Then we can define a special bilinear supply rate of the form $s(u, y) = \langle y, u \rangle = y^T u$ and this leads us to the definition of passivity.

Definition 1.3. *The system (1.32) with $m = p$ is said to be passive if it is dissipative with supply rate $s(u, y) = \langle y, u \rangle$. Furthermore, we say that (1.32) is strictly input (output) passive if it is dissipative with supply rate $s(u, y) = \langle y, u \rangle - \alpha \|u\|^2$ ($s(u, y) = \langle y, u \rangle - \beta \|y\|^2$) for a suitable α, $\beta > 0$. A passive system that is lossless is also called conservative.*

Taking for the electromechanical system (1.27) of Fig. 1.4 $u^T = [U_0, F_{ext}]$ as the plant input and $y^T = [i_L, v]$ as the plant output, we immediately see from (1.30) that the system (1.27) is passive. Moreover, since $p_{diss} = dv^2 + Ri_L^2 \geq \beta \|y\|^2$ with $0 < \beta < \min(d, R)$, (1.27) is also strictly output passive.

Remark 1.7. From the definition of passivity one can immediately see the connection between passivity and Lyapunov stability. Let us assume that (1.32) has an equilibrium at the origin, i.e., $f(0, 0) = 0$ and $h(0, 0) = 0$. Then, if (1.32) is passive with a positive definite C^1 storage function $V(x)$, $V(0) = 0$, the equilibrium $\bar{x} = 0$ of the free system, i.e. for $u = 0$, is stable in the sense of Lyapunov due to (1.36) and Theorem 1.1.

Passive systems have the pleasing property that the parallel and feedback interconnection of passive systems as shown in Fig. 1.5 is again passive. Suppose two passive systems of the form (1.32) with the states x_1, x_2, the inputs u_1, u_2 and the outputs y_1, y_2. Then there exist two storage functions $V_1(x_1)$ and $V_2(x_2)$ such that

$$V_1(x_1(t)) - V_1(x_1(t_0)) \leq \int_{t_0}^{t} \langle y_1, u_1 \rangle \, d\tau$$

$$V_2(x_2(t)) - V_2(x_2(t_0)) \leq \int_{t_0}^{t} \langle y_2, u_2 \rangle \, d\tau \ . \tag{1.37}$$

Substituting $u_1 = u_2 = u$ and $y = y_1 + y_2$ for the parallel connection, we get

$$V(x(t)) - V(x(t_0)) \leq \int_{t_0}^{t} \langle y, u \rangle \, d\tau \tag{1.38}$$

with $V = V_1 + V_2$ as the storage function of the parallel interconnected system and $x^T = [x_1^T, x_2^T]$. In this way, by inserting the connection conditions for the feedback connection $u_1 = e_1 - y_2$ and $u_2 = y_1 + e_2$ into (1.37), we obtain the inequality

$$V\left(x\left(t\right)\right) - V\left(x\left(t_0\right)\right) \leq \int_{t_0}^{t} \langle y_1, e_1 \rangle + \langle y_2, e_2 \rangle \, d\tau. \tag{1.39}$$

This shows the passivity of the closed-loop system with the input (e_1, e_2), the output (y_1, y_2), the state $x^T = \left[x_1^T, x_2^T\right]$ and the closed-loop storage function $V = V_1 + V_2$.

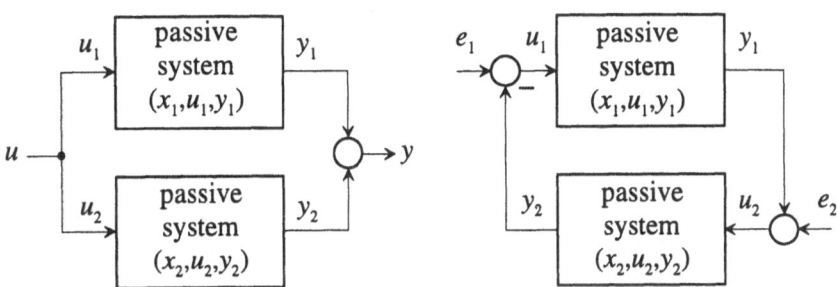

Fig. 1.5. Parallel and feedback interconnection of passive systems.

Subsequently, we will give necessary conditions for a non-linear system to be passive. For the sake of simplicity we will restrict our considerations to a special class of non-linear systems, which are affine in the input without feedthrough term

$$\frac{d}{dt}x = f\left(x\right) + \sum_{j=1}^{m} g_j\left(x\right) u_j \tag{1.40}$$
$$y = h\left(x\right)$$

with the smooth vector fields $f, g_1, \ldots, g_m$, the smooth functions $h_1, \ldots, h_m$, the state $x \in \mathcal{X} \subset R^n$, the control input $u \in \mathcal{U} \subset R^m$ and the output $y \in \mathcal{Y} \subset R^m$. Let us further assume that the origin is an equilibrium of (1.40), i.e., $f\left(0\right) = 0$ and $h\left(0\right) = 0$. Following [17] (see also [126]), we may state the following theorem without proof.

Theorem 1.4. *The necessary conditions for the system (1.40) to be passive with a C^2 storage function $V\left(x\right)$, $V\left(0\right) = 0$, are that the system (1.40)*

(1) has relative degree $\{1, \ldots, 1\}$ at $x = 0$ and
(2) is weakly minimum phase.

Remember that the system (1.40) has relative degree $\{r_1, \ldots, r_m\}$ at the point $\bar{x}$ if the following two conditions

(1) $L_{g_j} L_f^k h_i\left(x\right) = 0$ for all $1 \leq j, i \leq m$, all $0 \leq k < r_i - 1$ and all x in a neighborhood of $\bar{x}$ and

(2) the so-called decoupling matrix $\left[L_{g_j} L_f^{r_i-1} h_i\left(x\right)\right]$, $1 \leq j, i \leq m$ is non-singular at $x = \bar{x}$

are satisfied. If the system (1.40) has relative degree $\{1, \ldots, 1\}$ at $x = 0$ and the distribution spanned by the vector fields $\{g_1\left(x\right), \ldots, g_m\left(x\right)\}$ is involutive, it is always possible to find a local coordinate transformation in a neighborhood of $x = 0$ with the new coordinates (y, z) such that (1.40) takes the form

$$\frac{\mathrm{d}}{\mathrm{d}t}y = f_1\left(y, z\right) + \sum_{j=1}^{m} g_{1,j}\left(y, z\right) u_j$$
$$\frac{\mathrm{d}}{\mathrm{d}t}z = f_2\left(y, z\right) . \tag{1.41}$$

Recall that the so-called zero dynamics of (1.40) represent the internal dynamics of (1.40), when the output $y = h\left(x\right)$ is constrained to be identically zero. In the transformed system (1.41) the zero dynamics are given by the differential equation

$$\frac{\mathrm{d}}{\mathrm{d}t}z = f_2\left(0, z\right) = \bar{f}\left(z\right) . \tag{1.42}$$

Now, the system (1.40) is said to be weakly minimum phase if for the zero dynamics (1.42) there exists a positive definite C^2 function $W\left(z\right)$, $W\left(0\right) = 0$, such that $L_{\bar{f}} W \leq 0$ for all z in a neighborhood of $z = 0$. Concerning the notions of zero dynamics and relative degree the reader is referred to e.g., [52], [103], [144].

Remark 1.8. For linear SISO-systems without feedthrough term the conditions (1) and (2) of Theorem 1.4 can be easily checked by means of the associated transfer function. That is (1) the difference between the degree of the denominator and numerator of the transfer function is 1 and (2) the roots of the numerator have non-positive real parts and every root having a zero real part is simple.

Remark 1.9. The conditions (1) and (2) of Theorem 1.4 are even necessary and sufficient for the system (1.40) to be feedback equivalent to a passive system [17]. Thereby, the system (1.40) is said to be feedback equivalent to a passive system if there exists a state feedback law for (1.40) such that the closed-loop is passive. For more details on this topic, see [17], [126], [143].

1.2.4 Positive Realness

In this subsection, we will present the famous Kalman-Yakubovich-Popov lemma which gives a connection of the passivity property of a linear system with the existence of a solution of a set of algebraic equations containing the Lyapunov equation. However, before formulating this lemma for linear systems, we will derive an algebraic criterion due to [42] and [100], which can be understood as an extension of the Kalman-Yakubovich-Popov conditions to non-linear affine-input systems.

Non-linear Affine-input Systems. Let us at first concentrate on non-linear affine-input systems (1.40) but now with a throughput term $J(x)$

$$\frac{d}{dt}x = f(x) + \sum_{j=1}^{m} g_j(x) u_j \qquad (1.43)$$
$$y = h(x) + J(x) u .$$

We impose the assumption on all storage functions $V(x)$ including the available storage $V_a(x)$ that, whenever they exist, they are continuously differentiable. Hence we may apply Remark 1.5.

Definition 1.4. *The system (1.43) is said to be positive real if for all admissible u and all $t \geq t_0$ the inequality*

$$\int_{t_0}^{t} \langle y, u \rangle \, d\tau \geq 0 \qquad (1.44)$$

is satisfied for $x(t_0) = 0$.

In the case where $\langle y, u \rangle$ is the power which is externally supplied to the system (e.g., associated voltages and currents in an electrical network or associated generalized forces and velocities in a mechanical system), we see that the inequality condition (1.44) states that from the time t_0 more energy is flowing into the system than out of the system for all times $t > t_0$.

Remark 1.10. The positive real condition of (1.44) coincides with the definition of the passivity of an input-output system with zero-bias (see, e.g., [143], [144]).

It is apparent that there is a strong relation between positive real and passive systems. Before formulating a lemma due to [145] (see also [17], [143]), which clarifies this relation, we need the following definition.

Definition 1.5. *The system (1.32) is said to be reachable from $x(t_0) = 0$ if for every state $x \in \mathcal{X}$ there exists a time $t > 0$ and an admissible input $u \in \mathcal{U}$ such that $x = \varphi(t, 0, u(t))$.*

Lemma 1.1. *[145] Assume the system (1.43) is reachable from $x(t_0) = 0$. Then the system is passive if and only if it is positive real. Furthermore, the available storage $V_a(x)$ satisfies the relation $V_a(0) = 0$.*

The proof is omitted here but can be found in [145] or a newer version in [143].

In [42] and [100] an algebraic criterion, which is the generalization of the well-known Kalman-Yakubovich-Popov conditions to non-linear affine-input systems (1.43), is presented. At first let us recall the definition of a system being zero-state observable.

Definition 1.6. *The system (1.32) is said to be zero-state observable if for any trajectory with $u(t) = 0$ and $y(t) = 0$ for all $t \geq t_0$ implies $x(t) = 0$ for all $t \geq t_0$.*

Theorem 1.5. *[42], [100] Let the system (1.43) be reachable from 0 and zero-state observable. Then the system (1.43) is positive real if and only if there exist real-valued functions $l(x)$, $W(x)$ and a positive definite function $V(x)$ such that the following relations*

$$
\begin{aligned}
L_f V(x) &= -l^T(x)\, l(x) \\
\tfrac{1}{2}[L_{g_1} V(x), \ldots, L_{g_m} V(x)] &= h^T(x) - l^T(x)\, W(x) \\
W^T(x)\, W(x) &= J^T(x) + J(x)
\end{aligned}
\tag{1.45}
$$

hold.

Proof. (if): We claim that suitable real-valued functions $l(x)$, $W(x)$ and a positive definite function $V(x)$ satisfy (1.45). Then the change of $V(x)$ along a trajectory of (1.43) is given by

$$
\frac{d}{dt} V = L_f V(x) + \sum_{j=1}^{m} L_{g_j} V(x)\, u_j
\tag{1.46}
$$

or after inserting the relations (1.45) into (1.46)

$$
\frac{d}{dt} V = -\left(l(x) + W(x)\, u \right)^T \left(l(x) + W(x)\, u \right) + 2 \langle y, u \rangle \;.
\tag{1.47}
$$

Obviously,

$$
\frac{1}{2} \frac{d}{dt} V \leq \langle y, u \rangle
\tag{1.48}
$$

and hence for all admissible u, all initial values $x(t_0)$ and all $t \geq t_0$ we have

$$
\frac{1}{2} V(x(t)) - \frac{1}{2} V(x(t_0)) \leq \int_{t_0}^{t} \langle y, u \rangle \, dt,
\tag{1.49}
$$

which proves the passivity and due to Lemma 1.1 also positive realness.

(only if): We claim that (1.43) is positive real. Since (1.43) is assumed to be reachable from $x(t_0) = 0$, we may conclude from Lemma 1.1 that (1.43) is passive and the available storage $V_a(x)$ satisfies the relation $V_a(x(t_0)) = 0$. Therefore,

$$
0 \leq V_a(x(t)) - V_a(x(t_0)) \leq 2 \int_{t_0}^{t} \langle y, u \rangle \, dt
\tag{1.50}
$$

and due to our assumption that all storage functions are continuously differentiable, we get

$$0 \leq 2 \langle y, u \rangle - \frac{\mathrm{d}}{\mathrm{d}t} V_a \tag{1.51}$$

or

$$d(x, u) = 2 \langle y, u \rangle - \frac{\mathrm{d}}{\mathrm{d}t} V_a \geq 0 \tag{1.52}$$

with a positive semi-definite $d(x, u)$. It is worth mentioning that the factor 2 in (1.50) is introduced to obtain the equations in a form which fit the well-known Kalman-Yakubovich-Popov conditions in the linear case. The evaluation of $\frac{\mathrm{d}}{\mathrm{d}t} V_a$ along an integral curve of (1.43) yields

$$d(x, u) = 2h^T(x) u + 2u^T J^T(x) u - L_f V_a(x) - \sum_{j=1}^{m} L_{g_j} V_a(x) u_j. \tag{1.53}$$

Obviously, the quadratic term in u must be positive semi-definite and hence we may factor $J^T(x) + J(x) = W^T(x) W(x)$. By completing the squares, which is, of course, far from being unique, we choose $d(x, u)$ in the form

$$d(x, u) = (l(x) + W(x) u)^T (l(x) + W(x) u) \geq 0 . \tag{1.54}$$

Equation (1.53) together with (1.54) immediately brings the result (1.45).

Moreover, since (1.43) is zero-state observable all solutions $V(x)$ of (1.45) are positive definite (see, e.g., [42] or [143]). ∎

Linear Systems. Consider a linear time-invariant state-space system

$$\begin{aligned} \frac{\mathrm{d}}{\mathrm{d}t} x &= Ax + Bu \\ y &= Cx + Du \end{aligned} \tag{1.55}$$

with (A, B) reachable and (C, A) observable. From Theorem 1.5 we know that if (1.55) is positive real we find a positive definite solution $V(x) = x^T P x$, $P > 0$ and matrices L and W such that (1.45) is fulfilled. This leads to the well-known Kalman-Yakubovich-Popov conditions in the linear case

$$\begin{aligned} PA + A^T P &= -L^T L \\ PB &= C^T - L^T W \\ W^T W &= D^T + D . \end{aligned} \tag{1.56}$$

The positive realness of the linear system (1.55) can also be formulated in the frequency domain as a property of the associated transfer function (see, e.g., [59], [145])

$$Z(s) = C(sI - A)^{-1} B + D . \tag{1.57}$$

Definition 1.7. *A quadratic matrix $Z(s)$ of functions in a complex variable s is called positive real if for $\mathrm{Re}(s) > 0$ the following conditions*

(1) all elements of $Z(s)$ are analytic,
(2) $Z(s^*) = Z^*(s)$ and
(3) $Z^T(s^*) + Z(s)$ is positive semi-definite

*hold, where * denotes the complex conjugation. Moreover, we say $Z(s)$ is strictly positive real if $Z(s - \varepsilon)$ is positive real for some suitable $\varepsilon > 0$.*

For a transfer function $Z(s)$ with the property $\det(D) \neq 0$ we have the following lemma for characterizing strict positive realness [59], [144].

Lemma 1.2. *The transfer function $Z(s)$ of (1.57) is strictly positive real if and only if the conditions*

(1) A is Hurwitz and
(2) $\inf_{\omega \in R} \lambda_{\min}\left(Z(j\omega) + Z^T(-j\omega)\right) > 0$ with $\lambda_{\min}$ as the smallest eigenvalue of $Z + Z^T$

hold.

Remark 1.11. For the single-input single-output case, condition (2) of Lemma 1.2 simplifies to $\mathrm{Re}(Z(j\omega)) > 0$ for all ω and hence can be easily graphically checked by means of the Nyquist plot.

Furthermore, if $Z(s)$ is strictly positive real it is always possible to find an $\varepsilon > 0$ small enough such that $Z\left(s - \frac{\varepsilon}{2}\right) = C(sI - A_\varepsilon)^{-1}B + D$ is positive real and $A_\varepsilon = A + \frac{\varepsilon}{2}I$ is Hurwitz. Then the Kalman-Yakubovich-Popov conditions (1.56) can be rewritten for A_ε and the first equation of (1.56) becomes

$$PA + A^T P = -L^T L - \epsilon P . \tag{1.58}$$

Now, Theorem 1.5 together with (1.56) and (1.58) directly leads us to the celebrated Kalman-Yakubovich-Popov lemma (see, e.g., [59], [144]).

Theorem 1.6. *Consider the linear system (1.55) where the pair (A, B) is reachable and the pair (C, A) is observable. Then $Z(s)$ of (1.57) is strictly positive real if and only if there exist matrices L, W, a positive definite matrix P and an $\epsilon > 0$ such that the following relations*

$$\begin{aligned}
PA + A^T P &= -L^T L - \epsilon P \\
PB \quad &= C^T - L^T W \\
W^T W \quad &= D^T + D .
\end{aligned} \tag{1.59}$$

hold.

1.3 Absolute Stability and the Popov Criterion

1.3.1 Physical Observation IV

Let us consider again the simple mechanical system (1.9) of Fig. 1.2. We have already shown in Subsection 1.1.3 that the equilibrium $\bar{x} = 0$ is asymptotically stable. It is remarkable that in this example the stability does not rely on the particular form of the restoring force $F_c = \psi_F(z)$, it rather depends on the property that $\psi_F(z)$ satisfies the sector condition $k_1 z^2 \leq \psi_F(z) z \leq k_2 z^2$ with $0 < k_1 < k_2$.

A closer view shows that the mechanical system (1.9) can be represented as a feedback interconnection of a linear subsystem

$$
\begin{aligned}
\frac{d}{dt}z &= v \\
\frac{d}{dt}v &= -\frac{1}{m}(k_1 z + dv) + \frac{1}{m}u \\
y &= z
\end{aligned}
\tag{1.60}
$$

and a static non-linearity $\psi(z) = \psi_F(z) - k_1 z$, as shown in Fig. 1.6. Now, it is appropriate to pose the question under which conditions the equilibrium of such a feedback connected system according to Fig. 1.6 is asymptotically stable, if only the sector condition of the non-linearity is given, but not the non-linearity itself. The answer to this question leads to the notion of absolute stability.

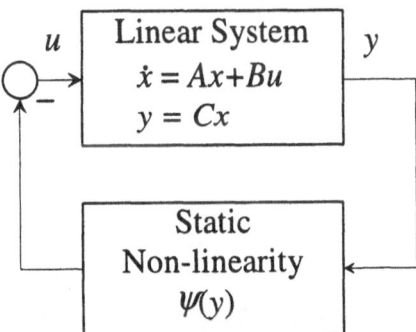

Fig. 1.6. Feedback interconnection of a linear system and a static non-linearity.

1.3.2 Mathematical Formulation IV: The Notion of Absolute Stability and the Popov Criterion

Given a feedback connected system, as shown in Fig. 1.6, where the linear time-invariant subsystem is described by

$$\frac{\mathrm{d}}{\mathrm{d}t}x = Ax + Bu$$
$$y \quad = Cx \tag{1.61}$$

with the state $x \in R^n$, the input $u \in R^m$ and the output $y \in R^m$. The non-linearity in the feedback path

$$u = -\psi(y) \tag{1.62}$$

is assumed to be memoryless, time-invariant and decentral, that is $\psi_i(y)$ only depends on y_i, $i = 1, \dots, m$. Further, the non-linearity is required to satisfy the sector condition

$$0 \le \psi^T(y)\,y \le y^T K y \tag{1.63}$$

with $K = \mathrm{diag}\,(k_i, i = 1, \dots, m)$, $k_i > 0$. Note that the lower bound 0 of the sector condition (1.63) can always be obtained by a suitable sector transformation.

Definition 1.8. *The feedback connected system (1.61) and (1.62) with a given sector condition (1.63) is said to be absolutely stable if the origin is globally asymptotically stable for any non-linearity belonging to the sector (1.63).*

The multivariable Popov criterion, stated in the next theorem, gives sufficient conditions for the absolute stability formulated as a strict positive realness property of a certain transfer matrix in the frequency domain (see, e.g., [59], [144] for further details).

Theorem 1.7. *Consider the feedback connected system (1.61) and (1.62), where A is Hurwitz, the pair (A, B) is reachable and the pair (C, A) is observable. Then the system is absolutely stable if there exists an $\eta \ge 0$ such that the transfer matrix*

$$Z_p(s) = I + (1 + \eta s)\,KC\,(sI - A)^{-1}\,B \tag{1.64}$$

is strictly positive real. Thereby, $-\frac{1}{\eta}$ is assumed not to coincide with an eigenvalue of A.

Proof. The main purpose of this proof is to point out the relation between the Popov criterion and the interconnection of passive systems (see also [143], [145]). Generally, the linear subsystem (1.61) is not passive. As one can see this is not even the case for the simple mechanical system (1.60). The underlying idea of the Popov criterion is that the system of Fig. 1.6, consisting of a linear subsystem and a sector non-linearity, can be represented as a feedback interconnection of two passive systems, as depicted in Fig. 1.7, with positive definite continuously differentiable storage functions.

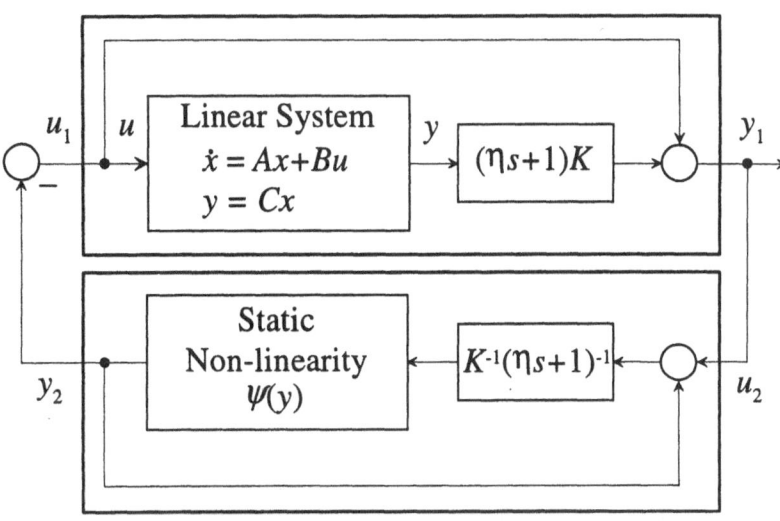

Fig. 1.7. The system of Fig. 1.6 redrawn as a feedback interconnection of two passive systems.

The transfer matrix of the system in the feedforward path of Fig. 1.7 with the input u_1 and the output y_1 is given by $Z_p(s)$ of (1.64) or equivalently, we may write [59]

$$Z_p(s) = (I + \eta KCB) + (KC + \eta KCA)(sI - A)^{-1}B . \qquad (1.65)$$

Thus the system

$$\frac{\mathrm{d}}{\mathrm{d}t}x_1 = Ax_1 + Bu_1$$
$$y_1 \quad = (KC + \eta KCA)x_1 + (I + \eta KCB)u_1 \qquad (1.66)$$

is a state-space representation of (1.65). From the fact that (C, A) is assumed to be observable and $-\frac{1}{\eta}$ is not an eigenvalue of A, we can deduce that $(KC + \eta KCA, A)$ is also observable. Since (A, B) is reachable, $(KC + \eta KCA, A)$ is observable and $Z_p(s)$ is strictly positive real, we may apply the Kalman-Yakubovich-Popov lemma of Theorem 1.6. With it, we find a positive definite storage function $V_1(x_1) = \frac{1}{2}x_1^T P x_1$, $P > 0$ such that

$$\frac{\mathrm{d}}{\mathrm{d}t}V_1 = \langle y_1, u_1 \rangle - \frac{1}{2}\epsilon x_1^T P x_1 - \frac{1}{2}(Lx_1 + Wu_1)^T (Lx_1 + Wu_1) \le \langle y_1, u_1 \rangle , \qquad (1.67)$$

which, of course, also says that the system (1.66) is passive.

The state-space representation of the system in the feedback path of Fig. 1.7 with the input u_2 and the output y_2 reads as

$$\eta \frac{d}{dt} x_2 = -x_2 + K^{-1} \left(u_2 + \psi \left(x_2 \right) \right)$$
$$y_2 \quad = \psi \left(x_2 \right) .$$

(1.68)

By means of the positive definite Lur'e type storage function

$$V_2 \left(x_2 \right) = \eta \sum_{i=1}^{m} k_i \int_0^{x_{2,i}} \psi_i \left(w \right) dw$$

(1.69)

and the sector condition (1.63), we immediately see with

$$\frac{d}{dt} V_2 = \langle y_2, u_2 \rangle - \psi^T \left(x_2 \right) \left(K x_2 - \psi \left(x_2 \right) \right) \le \langle y_2, u_2 \rangle$$

(1.70)

that (1.68) is passive. Thus, we have shown that Fig. 1.7 is in fact the interconnection of two passive systems. Thus, from Subsection 1.2.3 we also know that the closed-loop is passive with the positive definite storage function $V = V_1 + V_2$. Inserting the connection constraints $u_1 = -y_2$ and $u_2 = y_1$ into $\frac{d}{dt} V$, we obtain

$$\frac{d}{dt} V = -\frac{1}{2} \epsilon x_1^T P x_1 - \frac{1}{2} \left(L x_1 - W \psi \left(x_2 \right) \right)^T \left(L x_1 - W \psi \left(x_2 \right) \right)$$
$$- \psi^T \left(x_2 \right) \left(K x_2 - \psi \left(x_2 \right) \right)$$

(1.71)

which is clearly negative definite. This brings along the asymptotic stability of the origin.   ∎

Remark 1.12. In the single-input single-output case the test of the strict positive realness of $Z_p \left(s \right)$ from (1.64) simplifies to

$$\mathrm{Re} \left(Z \left(j\omega \right) \right) - \eta \omega \, \mathrm{Im} \left(Z \left(j\omega \right) \right) > -\frac{1}{k} \quad \text{for all } \omega \in R$$

(1.72)

with $Z \left(s \right) = c^T \left(sI - A \right)^{-1} b$. Note that here the non-linearity $\psi \left(y \right)$ satisfies the sector condition $0 \le \psi \left(y \right) y \le ky^2$, $k > 0$. The condition (1.72) can be easily graphically investigated by means of the so-called Popov plot, where $\mathrm{Re} \left(Z \left(j\omega \right) \right)$ is plot versus $\omega \, \mathrm{Im} \left(Z \left(j\omega \right) \right)$.

Remark 1.13. In connection with absolute stability we find, apart from the Popov criterion, the so-called circle criterion, which is also applicable for time-varying and non-decentral non-linearities. However, in principle these two criteria apply to two different classes of systems and therefore, a direct comparison is not necessarily meaningful. For more information and details the reader is referred to e.g., [59], [144].

1.4 PCH-systems and PCHD-systems

Undoubtedly, the energy balance of a physical system with the change in the energy storage, together with the energy in- and outflows plays a crucial role in the stability considerations. This is also reflected in the fact that the energy stored in a physical system is always suggested as an appropriate Lyapunov function candidate. Moreover, in recent decades the energy concept has again gained in importance since it turned out that it also serves as a basis for the design of non-linear controllers and not only as a tool for analyzing the stability. Today, the available design strategies based on dissipativity and passivity considerations are manifold and can be found in the literature under the headings non-linear H_∞-control, passivity-based control, energy shaping, damping injection etc.. A full overview of this approach can be found e.g., in [107], [126], [143] and the literature cited therein.

Due to this development it is understandable that we endeavour to formulate a generalized canonical description of passive systems which will give a clear insight into the energy situation of a passive system. We will henceforth confine the considered class of non-linear systems to the case with affine input. Then, we say a system is in passive canonical form if it can be expressed as

$$\frac{\mathrm{d}}{\mathrm{d}t}x = (J(x) - S(x))\left(\frac{\partial V}{\partial x}\right)^T + G_e(x)e \qquad (1.73)$$

with the state $x \in \mathcal{X} \subset R^n$, the external inputs $e^T = [u^T, d^T]$, consisting of the control input $u \in \mathcal{U} \subset R^m$ and the exogenous inputs (disturbances and reference inputs) $d \in \mathcal{D} \subset R^p$, the C^1 positive definite storage function $V(x)$, $V(0) = 0$, the matrices $G_e(x)$, $J(x) = -J^T(x)$ and $S(x) = S^T(x) \geq 0$ whose entries are smoothly depending on x. By choosing the output $y \in \mathcal{Y} \subset R^{m+p}$ in the form

$$y = G_e^T(x)\left(\frac{\partial V}{\partial x}\right)^T, \qquad (1.74)$$

one can immediately convince oneself that the system (1.73) and (1.74) is passive with the storage function $V(x)$, since the differential dissipation inequality is given by

$$\frac{\mathrm{d}}{\mathrm{d}t}V = \langle y, e\rangle - \left(\frac{\partial V}{\partial x}\right)S(x)\left(\frac{\partial V}{\partial x}\right)^T \leq \langle y, e\rangle . \qquad (1.75)$$

However, the representation of (1.73) and (1.74) is exactly what in [143] is called a port-controlled Hamiltonian system with dissipation (PCHD-system) or in [22], [103] a system in extended Hamiltonian canonical form. In order to be consistent with the literature, we will henceforth follow the terminology of [143] and call (1.73) and (1.74) a PCHD-system. Thereby, $J(x)$ represents the internal energy-preserving interconnection, $S(x)$ covers the dissipative effects and $G_e(x)$ describes the energy flows with the system environment via the system ports.

Throughout this work, we will be confronted with two special cases of (1.73). At first, if $S(x) = 0$, then (1.75) holds with equality and, following Definition 1.3, the system (1.73) is said to be lossless with respect to the supply rate $\langle y, e \rangle$, or conservative. Then, the system will be called a port-controlled Hamiltonian system or PCH-system for short [143]. Typical representatives of this class of system are well known from classical mechanics, namely the Hamiltonian systems, and they will be discussed later on in Chapter 4. The second case covers the situation where $\left(\frac{\partial V}{\partial x}\right) G_u(x)$ is generically zero with $G_e(x) e = G_u(x) u + G_d(x) d$. In the single-input single-output case $G_u(x)$ can be expressed in the form $G_u(x) = J_u(x) \left(\frac{\partial V}{\partial x}\right)^T$ with $J_u(x) = -J_u^T(x)$. Consequently, (1.73) can be written as

$$\frac{\mathrm{d}}{\mathrm{d}t} x = (J(x) + J_u(x) u - S(x)) \left(\frac{\partial V}{\partial x}\right)^T + G_d(x) d . \tag{1.76}$$

The physical interpretation of this case is that the control input u only influences the internal energy flow and thus does not change the total amount of energy stored in the system. It is quite obvious that in this connection the output, as defined in (1.74), is only meaningful for $G_e(x) = G_d(x)$. We will see in Chapters 3 and 5 that the Ćuk-converter, a special dc-to-dc power converter, and the pump-displacement-controlled rotational piston actuator are typical applications of the system (1.76).

It is worth mentioning that henceforth we do not intend to answer the question under which conditions a general mathematical model can be transformed into the form of a PCHD-system (1.73) and (1.74). But we have already seen in Theorem 1.4 of Subsection 1.2.3 that for a mathematical model with given output functions the necessary conditions for a system being passive are rather restrictive. Furthermore, in [17] it is proven that these conditions are even necessary and sufficient for the system being feedback equivalent to a passive system (see also Remark 1.9). Thus the reader is referred to e.g., [17], [126] and [143] to find an answer to this question.

Moreover, we claim that a physical system which is passive and which may be described by an explicit system of non-linear differential equations automatically induces a system description according to the passive canonical form of (1.73) and (1.74). It does not seem to be necessary to derive in a first step the mathematical model of a physical system and try to find the passivity structure in a second step. This is why, we focus our attention in the next chapters on the procedure: how the mathematical model of a physical system must be set up in order to directly obtain the associated passive canonical form.

Remark 1.14. In [133] a slightly different version of (1.73) is suggested as a passivity based control canonical form where the system is further extended by so-called locally destabilizing effects, which are taken into account by an additional symmetric, but positive semi-definite matrix $S_p(x)$.

Remark 1.15. To a certain extent it is also possible to generalize the passive canonical form of (1.73) and (1.74) to the infinite dimensional case. Here, $J(x)$ and $S(x)$ have to be replaced by a skew-adjoint and a self-adjoint non-negative differential operator, respectively and instead of the energy storage function an energy functional must be used. The generalization of this topic is beyond the scope of this work, although Chapter 4 presents some analogies of finite and infinite-dimensional mechanical systems with application to smart structures. More details concerning the theoretical background of the passive canonical form in the infinite-dimensional case can be found, e.g. in [82].

2. Some Non-linear Control Design Strategies

During the last decades there have been some significant advances in the area of non-linear control system design from both the theoretical and the practical point of view. The research effort on non-linear control is important because of the more demanding performance required in practical applications and because most of the physical systems are non-linear in nature. Moreover, the increasing availability of low cost digital processors and the increasing power of computer programs for numeric and symbolic computation make the practical use of these non-linear control strategies possible.

In this chapter we briefly summarize the results of those non-linear model-based control approaches which, on the one hand will be used in the subsequent chapters and which, on the other hand have a more or less strong relation to the dissipativity and passivity concept. Essentially, these are the non-linear state feedback H_2-design for affine-input systems with and without integral term (see, e.g., [126], [143]), the non-linear state feedback H_∞-design for affine input systems (see, e.g., [53], [58], [143]) and the passivity-based control concept (see, e.g., [107], [143]). As the reader may rightly notice, these three control synthesis tools discussed within this chapter are just a very small part of the full range available in literature. This should not give the impression to the reader that these non-linear control strategies are always our first choice for practical applications. The main intention of the work is rather to elaborate the physical structure of electrical, electromechanical, mechanical and hydraulic systems, which will be done in detail in Chapters 3, 4 and 5. The choice of the right control design strategy for a successful practical implementation depends strongly on the considered application, the demands on the closed-loop, the restrictions and limitations of the plant itself and the actuators, the measurable quantities and their accuracy, the limitation of the real-time hard- and software platform etc. and cannot be answered generally and without preceding detailed investigation of the plant to be controlled. Without laying claim to completeness, the reader is referred to the following excellent textbooks and contributions to get a more complete idea of the different non-linear control design strategies, that are e.g., [52], [91], [103], [144] for differential geometric methods, e.g., [60], [91], [126] for backstepping and adaptive non-linear design strategies, e.g., [58], [143], for non-linear H_∞-design, e.g., [115] for the differential algebraic approach, e.g., [28] for the flatness concept and e.g., [107], [126], [143] for passivity-based control. However, all the model-based non-linear control approaches have one fact in common, namely that somehow the knowledge of the underlying physical structure helps to solve the design problems. In order to stress this argument let us mention as examples the solution of the Hamilton-Jacobi-Bellman-Isaacs equation or inequality for the non-linear H_∞-design; the solution of the Frobenius-type partial differential equations for the input-state or input-output exact linearization; the determination of the flat outputs within the flatness approach, or the suitable choice of the Lyapunov-like function within the backstepping or composite Lyapunov design.

2.1 Non-linear State Feedback H_2-design

Consider an affine-input system

$$
\begin{aligned}
\frac{\mathrm{d}}{\mathrm{d}t}x &= f\left(x\right) + \sum_{j=1}^{m} g_j\left(x\right) u_j \\
y &= h\left(x\right)
\end{aligned}
\tag{2.1}
$$

with the smooth vector fields $f, g_1, \ldots, g_m$, the smooth functions $h_1, \ldots, h_q$, the state $x \in \mathcal{X} \subset R^n$, the control input $u \in \mathcal{U} \subset R^m$ and the output $y \in \mathcal{Y} \subset R^q$. Let us assume without restriction of generality that the origin is an equilibrium of (2.1) for $u = 0$, i.e., $f\left(0\right) = 0$ and $h\left(0\right) = 0$.

Now, the goal of the non-linear H_2-design is to find a control law

$$
u = u(x), \qquad u(0) = 0
\tag{2.2}
$$

such that the origin is rendered asymptotically stable and the objective function

$$
J_2 = \inf_{u \in L_2^m[0,\infty)} \int_0^\infty \Lambda_2\left(x, u\right) \mathrm{d}t
\tag{2.3}
$$

with

$$
\Lambda_2\left(x, u\right) = \frac{1}{2}\left(\beta \left\|y\right\|^2 + \left\|u\right\|^2\right), \beta > 0
\tag{2.4}
$$

is minimized with respect to u. Here and subsequently the integral is always evaluated along a solution of (2.1). It is worth mentioning that $\Lambda_2\left(x, u\left(x\right)\right)$ is in any case a positive semi-definite function of x and if, furthermore, a certain observability condition is satisfied, $\Lambda_2\left(x, u\left(x\right)\right)$ is even positive definite. Let us assume there exists a real-valued C^1 function $V\left(x\right)$, such that the relation

$$
V\left(x\right) = \inf_u \int_{t_0}^\infty \Lambda_2\left(x, u\right) \mathrm{d}\tau, \quad x = x\left(t_0\right)
\tag{2.5}
$$

is met. From the principle of optimality (see, e.g., [5], [20]) we know that (2.5) is equivalent to

$$V\left(x\left(t_0\right)\right) = \inf_{u} \left(\int_{t_0}^{t} \Lambda_2\left(x, u\right) d\tau + \inf_{\tilde{u}} \int_{t}^{\infty} \Lambda_2\left(x, \tilde{u}\right) d\tau \right) \tag{2.6}$$

or

$$V\left(x\left(t_0\right)\right) = \inf_{u} \left(\int_{t_0}^{t} \Lambda_2\left(x, u\right) d\tau + V\left(x\left(t\right)\right) \right) . \tag{2.7}$$

Thus, from (2.7) it follows that V must satisfy the well known Hamilton-Jacobi-Bellman equation (HJBe)

$$\min_{u} \left(L_f V\left(x\right) + \sum_{j=1}^{m} L_{g_j} V\left(x\right) u_j + \Lambda_2\left(x, u\right) \right) = 0 . \tag{2.8}$$

By substituting $\Lambda_2\left(x, u\right)$ from (2.4) into (2.8), we get the optimal choice u^* of u

$$u_j^* = -L_{g_j} V, \quad j = 1, \dots, m . \tag{2.9}$$

Equation (2.8) together with (2.9) yields the HJBe in the form

$$L_f V + \frac{1}{2} \left(\beta \left\| y \right\|^2 - \sum_{j=1}^{m} \left(L_{g_j} V\right)^2 \right) = 0 . \tag{2.10}$$

Generally, for practical applications it is difficult, if not impossible, to find a solution of the HJBe (2.10), see also Remark 2.1. Therefore, we are often satisfied to solve the so-called Hamilton-Jacobi-Bellman inequality (HJBi)

$$L_f V + \frac{1}{2} \left(\beta \left\| y \right\|^2 - \sum_{j=1}^{m} \left(L_{g_j} V\right)^2 \right) \leq 0, \tag{2.11}$$

which, of course, only leads to a suboptimal solution of the non-linear H_2-design problem.

Theorem 2.1. *Given the affine-input system (2.1) and suppose $V\left(x\right)$, $V\left(0\right) = 0$, is a C^1 positive definite solution of the HJBe (2.10). Then, if the system*

$$\begin{aligned} \frac{d}{dt} x &= f\left(x\right) \\ \bar{y}^T &= [h_1, \dots, h_q, L_{g_1} V, \dots, L_{g_m} V] \end{aligned} \tag{2.12}$$

is zero-state observable, the state feedback (2.9) solves the optimal non-linear H_2-design problem. Furthermore, if $V\left(x\right)$, $V\left(0\right) = 0$, is a C^1 positive definite solution of the HJBi (2.11), the state feedback (2.9) is only suboptimal.

Proof. The closed-loop of (2.1) and (2.9) reads as

$$\frac{\mathrm{d}}{\mathrm{d}t}x = f(x) - \sum_{j=1}^{m} g_j(x) L_{g_j} V(x)$$
$$y = h(x) .$$

(2.13)

Obviously, $V(x)$ as the positive definite solution of the HJBe (2.10) or HJBi (2.11) serves as an appropriate Lyapunov function candidate. By calculating the change of V along an integral curve of (2.13), we get with (2.10) (with (2.11))

$$\frac{\mathrm{d}}{\mathrm{d}t}V = L_f V - \sum_{j=1}^{m} (L_{g_j} V)^2 \underset{(\le)}{=} -\frac{1}{2}\left(\beta \|h\|^2 + \sum_{j=1}^{m} (L_{g_j} V)^2\right) \le 0 .$$

(2.14)

Since $\frac{\mathrm{d}}{\mathrm{d}t}V$ is negative semi-definite we may conclude that the origin is stable in the sense of Lyapunov. But the zero-state observability of (2.12) says that for any trajectory, $\bar{y}_i(t) = 0, i = 1, \ldots, q+m$ for all $t \ge t_0$ implies $x(t) = 0$ for all $t \ge t_0$ (see Definition 1.6). Hence, by means of LaSalle's invariance principle of Theorem 1.2 the origin is asymptotically stable and $\lim_{t \to \infty} x(t) = 0$.

In the next step, we will show that if V is a solution of the HJBe (2.10) the state feedback (2.9) in fact minimizes the objective function (2.3). We will prove this by contradiction as it is done in [126]. Let us assume that (2.9) is not the optimizing control law, instead the optimal state feedback reads as

$$u_j^* = -L_{g_j} V + v_j, \quad j = 1, \ldots, m$$

(2.15)

for some suitable v_j. Then, by inserting (2.15) into the objective function (2.3), we get

$$J_2 = \int_0^\infty \frac{1}{2}\left(\sum_{j=1}^{m} (L_{g_j} V)^2 - 2\sum_{j=1}^{m} v_j L_{g_j} V + \|v\|^2 + \beta \|y\|^2\right) \mathrm{d}t$$

(2.16)

or with (2.10)

$$J_2 = \int_0^\infty \left(-L_f V - \sum_{j=1}^{m} L_{g_j} V (v_j - L_{g_j} V) + \frac{1}{2}\|v\|^2\right) \mathrm{d}t$$
$$= \int_0^\infty \left(-\frac{\mathrm{d}}{\mathrm{d}t}V + \frac{1}{2}\|v\|^2\right) \mathrm{d}t$$
$$= -\lim_{t \to \infty} V(x(t)) + V(x(0)) + \frac{1}{2}\int_0^\infty \|v\|^2 \mathrm{d}t .$$

(2.17)

Since the origin is asymptotically stable, it follows that $\lim_{t \to \infty} V(x(t)) = 0$ and hence J_2 is obviously minimal if v is identically zero. This contradicts the assumption that (2.9) is not the optimizing control law.   ∎

For linear systems

$$\frac{\mathrm{d}}{\mathrm{d}t}x = Ax + Bu$$
$$y = Cx$$

(2.18)

the HJBe (2.10) simplifies for $\beta = 1$ with $V = \frac{1}{2}x^T P x$, $P > 0$, to the well known algebraic Riccati equation

$$x^T \left(PA + A^T P + C^T C - PBB^T P\right) x = 0 \, ,$$

(2.19)

which has indeed a unique positive definite solution P if the pair (A, B) is reachable and the pair (C, A) observable. Note that the observability condition of the system (2.12) in Theorem 2.1 also implies the reachability of (A, B) and the observability of (C, A). Consequently, the HJBi (2.11) for linear systems yields

$$x^T \left(PA + A^T P + C^T C - PBB^T P\right) x \leq 0 \, .$$

(2.20)

Remark 2.1. It is worth mentioning that in the literature also procedures for obtaining an approximate solution of the HJBe (2.10) are suggested, see, e.g., [142]. Thereby, the solution V of the HJBe (2.10) is gradually constructed, starting with a solution of the algebraic Riccati equation (2.19) of the associated linearized system. The drawback of these approximate solutions is that they are only valid in a small neighborhood of the equilibrium.

Next, let us consider a PCHD-system (see (1.73), (1.74) of Chapter 1)

$$\frac{\mathrm{d}}{\mathrm{d}t}x = (J(x) - S(x)) \left(\frac{\partial V}{\partial x}\right)^T + G_u(x) u$$
$$y = G_u^T(x) \left(\frac{\partial V}{\partial x}\right)^T \, .$$

(2.21)

Then it is easy to see that the storage function $V(x)$ is a solution of the HJBi (2.11) for $\beta = 1$, because

$$-\left(\frac{\partial V}{\partial x}\right) S(x) \left(\frac{\partial V}{\partial x}\right)^T \leq 0 \, .$$

(2.22)

Furthermore, in the special case of a lossless system, i.e. $S(x) = 0$, the storage function $V(x)$ even solves the HJBe (2.10), see also [143] and for a detailed investigation for infinite-dimensional Hamiltonian systems Chapter 4, in particular, Proposition 4.1. Another pleasing result can be obtained for the case, when the control input u only influences the internal energy flow (compare also with (1.76) of Chapter 1). Then the associated PCHD-like system reads as

$$\frac{\mathrm{d}}{\mathrm{d}t}x = \left(J\left(x\right) + J_u\left(x\right)u - S\left(x\right)\right)\left(\frac{\partial V}{\partial x}\right)^T$$
$$y \quad = G_y^T\left(x\right)\left(\frac{\partial V}{\partial x}\right)^T . \tag{2.23}$$

We can immediately observe that with the storage function $V\left(x\right)$ the HJBi (2.11) for $\beta = 1$ takes the form

$$-\left(\frac{\partial V}{\partial x}\right)\left(S\left(x\right) - \frac{1}{2}G_y\left(x\right)G_y^T\left(x\right)\right)\left(\frac{\partial V}{\partial x}\right)^T \leq 0 . \tag{2.24}$$

Clearly, $u = 0$ solves the suboptimal non-linear H_2-design problem if $S\left(x\right) - \frac{1}{2}G_y^T\left(x\right)G_y\left(x\right) \geq 0$ and if the associated observability condition is fulfilled.

Remark 2.2. In [126] a more general objective function (2.3) of the form

$$J_2 = \inf_{u \in L_2^m[0,\infty)} \int_0^\infty \left(l\left(x\right) + u^T R\left(x\right)u\right)\mathrm{d}t \tag{2.25}$$

with $l\left(x\right) \geq 0$ and $R\left(x\right) > 0$ is used for designing a feedback controller. However, the modifications of the HJBe (2.10), the HJBi (2.11) as well as of the optimal control law (2.9) can be obtained in a straightforward manner. The reader is referred to [126] for more details and many other aspects.

2.2 Non-linear State Feedback H_2-design with Integral Term

In many applications it is desirable that the controller contains an integral term in order to make the stationary error in the plant output zero and to compensate for parameter variations. But in general an integrator in the controller partially destroys the physical structure and hence makes a systematic controller design far more difficult. In the following we will restrict ourselves to a special type of non-linear SISO-systems of the form

$$\frac{\mathrm{d}}{\mathrm{d}t}x = Ax + b\left(x\right)u$$
$$y \quad = c^T x \tag{2.26}$$

with the plant input u and the plant output y. Further let us assume that the matrix A is Hurwitz and the pair (A, c) is observable.

Typical representatives of this class of system are systems of the type (1.76) where the skew-symmetric matrices J and J_u, the positive semi-definite matrix S and the matrix G_d have constant entries and the storage function $V\left(x\right)$ is quadratic, i.e. $V\left(x\right) = \frac{1}{2}x^T P_V x$, with the positive definite matrix P_V. Under these conditions (1.76) reads as

$$\frac{\mathrm{d}}{\mathrm{d}t} x = (J - S + J_u u) P_V x + G_d d \qquad (2.27)$$
$$y \quad = c^T x,$$

with the plant input u, the plant output y and the exogenous inputs d, which are assumed to be constant but unknown. Now, if for a given constant plant input $u = \bar{u}$ the matrix $(J - S + J_u \bar{u})$ is Hurwitz, the operating point $\bar{x} = -P_V^{-1} (J - S + J_u \bar{u})^{-1} G_d d$ is determined unambiguously by $\bar{u}$ and d. Then, by means of a simple change of coordinates $x = \bar{x} + \Delta x$ and $u = \bar{u} + \Delta u$ the operating point is shifted to the origin and the system (2.27) can be rewritten in the form

$$\frac{\mathrm{d}}{\mathrm{d}t} \Delta x = (J - S + J_u \bar{u}) P_V \Delta x + J_u P_V (\bar{x} + \Delta x) \Delta u \qquad (2.28)$$
$$\Delta y \quad = c^T \Delta x .$$

Obviously, (2.28) has the same structure as (2.26). We will see in Chapters 3 and 5 that the Ćuk-converter and the pump-displacement-controlled hydraulic rotational piston actuator are typical applications of the system (2.26).

Now, we are ready to formulate a proposition for the non-linear state feedback H_2-design with integral term:

Proposition 2.1. *Given the system (2.26) and suppose the matrix A is Hurwitz, the pair (A, c) is observable and the condition $c^T A^{-1} b(0) \neq 0$ is satisfied. Then the non-linear state feedback controller with integral term*

$$\frac{\mathrm{d}}{\mathrm{d}t} x_I = c^T x$$
$$u \quad = -x^T \left(P_{11} + (A^{-1})^T c c^T A^{-1} p_{22} \right) b(x) + p_{22} c^T A^{-1} b(x) x_I, \qquad (2.29)$$

with P_{11} as the unique positive definite solution of the Lyapunov equation

$$P_{11} A + A^T P_{11} + \beta c c^T = 0 \qquad (2.30)$$

solves the suboptimal non-linear H_2-design problem for the controller parameters β, $p_{22} > 0$. Furthermore, the parameters β and p_{22} can be used to adjust the closed-loop performance.

Proof. At first it is worth mentioning that the Lyapunov equation (2.30) has a unique positive definite solution since A is Hurwitz and the pair (A, c) observable, see, e.g., [144]. Next we will show that the C^1 positive definite function

$$V(x) = \frac{1}{2} x^T P_{11} x + \frac{1}{2} p_{22} \left(c^T A^{-1} x - x_I \right)^2 \qquad (2.31)$$

is a solution of the Hamilton-Jacobi-Bellman inequality (2.11) of the augmented system (2.26)

$$\frac{d}{dt}\begin{bmatrix} x \\ x_I \end{bmatrix} = \begin{bmatrix} A & 0 \\ c^T & 0 \end{bmatrix} \begin{bmatrix} x \\ x_I \end{bmatrix} + \begin{bmatrix} b(x) \\ 0 \end{bmatrix} u \ . \tag{2.32}$$

The associated optimal choice of the state feedback law (2.9) reads as

$$u(x, x_I) = -\frac{\partial V}{\partial x} b(x) = -x^T P_{11} b(x) - p_{22} \left(c^T A^{-1} x - x_I \right) c^T A^{-1} b(x)$$
$$= -x^T \left(P_{11} + \left(A^{-1} \right)^T c c^T A^{-1} p_{22} \right) b(x) + p_{22} c^T A^{-1} b(x) x_I \ . \tag{2.33}$$

By inserting $V(x)$ from (2.31) into the HJBi (2.11) and by using (2.30), we get

$$\tfrac{1}{2} x^T \left(P_{11} A + A^T P_{11} \right) x + p_{22} \left(c^T A^{-1} x - x_I \right) \left(c^T - c^T \right) x + \tfrac{1}{2} \beta \|y\|^2 -$$
$$\tfrac{1}{2} \left(\tfrac{\partial V}{\partial x} b(x) \right)^2 = -\tfrac{1}{2} \left(\tfrac{\partial V}{\partial x} b(x) \right)^2 = -\tfrac{1}{2} u(x, x_I)^2 \leq 0 \ . \tag{2.34}$$

Thus the HJBi is obviously satisfied. It remains to show that the system

$$\frac{d}{dt}\begin{bmatrix} x \\ x_I \end{bmatrix} = \begin{bmatrix} A & 0 \\ c^T & 0 \end{bmatrix} \begin{bmatrix} x \\ x_I \end{bmatrix}$$
$$\bar{y}^T \qquad = \left[c^T x, u(x, x_I) \right] \tag{2.35}$$

with $u(x, x_I)$ from (2.33) is zero-state observable. From $c^T x = 0$ we can conclude that $x = 0$ since (A, c) is observable. Furthermore, from $u(0, x_I) = p_{22} c^T A^{-1} b(0) x_I = 0$ it follows directly that $x_I = 0$ due to the assumption that $c^T A^{-1} b(0) \neq 0$. Hence the system (2.35) is indeed zero-state observable. Now, by Theorem 2.1 the state feedback law (2.33) solves the suboptimal non-linear H_2-design problem for the augmented system (2.32). This completes the proof. ∎

Remark 2.3. It is clear that the Lyapunov equation (2.30) of Proposition 2.1 has for every positive semi-definite matrix $Q = \Gamma^T \Gamma$ a unique positive definite solution P_{11}

$$P_{11} A + A^T P_{11} + Q = 0 \tag{2.36}$$

if the pair (A, Γ) is observable. This allows us to take into consideration other objective functions than (2.3). Furthermore, Proposition 2.1 can also be extended to the MIMO-case in a straightforward way when the number of control inputs u equals the number of outputs y.

2.3 Non-linear State Feedback H_∞-design

For the non-linear H_∞-design the affine-input system (2.1) is extended by the exogenous inputs (disturbances and reference inputs) $d \in \mathcal{D} \subset R^p$

$$\frac{\mathrm{d}}{\mathrm{d}t}x = f(x) + \sum_{j=1}^m g_j(x) u_j + \sum_{i=1}^p k_i(x) d_i$$
$$y = h(x) \tag{2.37}$$

with the smooth vector fields $f, g_1, \dots, g_m, k_1, \dots, k_p$, the smooth functions $h_1, \dots, h_q$, the state $x \in \mathcal{X} \subset R^n$, the control input $u \in \mathcal{U} \subset R^m$ and the output $y \in \mathcal{Y} \subset R^q$. Let us again assume without restriction of generality that the origin is an equilibrium of (2.37) for $u = 0$ and $d = 0$, i.e., $f(0) = 0$ and $h(0) = 0$.

The goal of the non-linear H_∞-design is to find a control law

$$u = u(x), \qquad u(0) = 0 \tag{2.38}$$

such that the objective function

$$J_\infty = \sup_{d \in L_2^p[0,\infty)} \inf_{u \in L_2^m[0,\infty)} \int_0^\infty \Lambda_\infty(x, u, d)\, \mathrm{d}t \tag{2.39}$$

with

$$\Lambda_\infty(x, u, d) = \frac{1}{2}\left(\beta \|y\|^2 + \|u\|^2 - \gamma \|d\|^2\right), \ \beta > 0 \tag{2.40}$$

is minimized with respect to u and maximized with respect to d, whereas the so-called disturbance attenuation level $\gamma > 0$ must be chosen such that the problem is solvable. In a second step, we also try to find the minimum value of γ. From the theory of differential games (see, e.g., [10], [80]) it follows that one has to look for a positive (semi)-definite solution $V(x)$ of the Hamilton-Jacobi-Bellman-Isaacs equation (HJBIe)

$$\max_d \min_u H_\gamma\left(x, \frac{\partial}{\partial x}V, u, d\right) = 0 \tag{2.41}$$

with the associated Hamiltonian function

$$H_\gamma\left(x, \frac{\partial}{\partial x}V, u, d\right) = L_f V(x) + \sum_{j=1}^m L_{g_j} V(x) u_j + \sum_{i=1}^p L_{k_i} V(x) d_i + \Lambda_\infty. \tag{2.42}$$

Now, in our case H_γ has a unique global saddle point with the property (see, e.g., [58], [141])

$$H_\gamma\left(x, \frac{\partial}{\partial x}V, u^*, d\right) \le H_\gamma\left(x, \frac{\partial}{\partial x}V, u^*, d^*\right) \le H_\gamma\left(x, \frac{\partial}{\partial x}V, u, d^*\right),$$

(2.43)

where u^* and d^* are determined respectively by

$$u_j^* = -L_{g_j}V, \, j = 1, \ldots, m \quad \text{and} \quad d_i^* = \frac{1}{\gamma}L_{k_i}V, \, i = 1, \ldots, p.$$

(2.44)

Thus, by inserting (2.44) into (2.41) the HJBIe reads as

$$L_fV + \frac{1}{2}\left(\beta\|y\|^2 - \sum_{j=1}^{m}(L_{g_j}V)^2 + \frac{1}{\gamma}\sum_{i=1}^{p}(L_{k_i}V)^2\right) = 0.$$

(2.45)

Analogous to the HJBe (2.10) the HJBIe is also rather difficult to solve and hence we are usually content with a solution of the Hamilton-Jacobi-Bellman-Isaacs inequality (HJBIi)

$$L_fV + \frac{1}{2}\left(\beta\|y\|^2 - \sum_{j=1}^{m}(L_{g_j}V)^2 + \frac{1}{\gamma}\sum_{i=1}^{p}(L_{k_i}V)^2\right) \le 0.$$

(2.46)

Theorem 2.2. *[141] Given the affine-input system (2.37). Then, if $V(x)$, $V(0) = 0$, is a positive semi-definite solution of the HJBIe (2.45) or the HJBIi (2.46), the state feedback*

$$u_j^* = -L_{g_j}V, \, j = 1, \ldots, m$$

(2.47)

renders the closed-loop dissipative with the supply rate

$$\frac{1}{2}\left(\gamma\|d\|^2 - \|u^*\|^2 - \beta\|y\|^2\right)$$

(2.48)

or equivalently the closed-loop has L_2-gain less equal γ from the input d to the output $z^T = \left[\sqrt{\beta}y^T, (u^*)^T\right]$. We then say (2.47) solves the suboptimal non-linear H_∞-design problem. If, furthermore, we find the smallest $\gamma^* \ge 0$ such that for all $\gamma > \gamma^*$ the state feedback (2.47) makes the closed-loop having L_2-gain less equal γ, then the non-linear H_∞-design problem is solved optimally.

Proof. [141] The closed-loop of (2.37) with the control law (2.47) takes the form

$$\frac{\mathrm{d}}{\mathrm{d}t}x = f(x) - \sum_{j=1}^{m}g_j(x)L_{g_j}V(x) + \sum_{i=1}^{p}k_i(x)d_i$$
$$y = h(x).$$

(2.49)

Suppose $V(x)$, $V(0) = 0$, is a positive semi-definite solution of the HJ-BIe (2.45) or the HJBIi (2.46). Then, by adding the expression $-\frac{1}{2}\|u^*\|^2 + \sum_{i=1}^p L_{k_i} V d_i$ to the left and right hand side of the HJBIe (2.45) (or the HJBIi (2.46)), we get

$$
L_f V - \sum_{j=1}^m \left(L_{g_j} V\right)^2 + \sum_{i=1}^p L_{k_i} V d_i \overset{=}{\underset{(\leq)}{}} \frac{1}{2}\left(\gamma\|d\|^2 - \|u^*\|^2 - \beta\|y\|^2\right)
$$
$$
- \frac{1}{2}\sum_{i=1}^p \left(\frac{1}{\sqrt{\gamma}}L_{k_i} V - \sqrt{\gamma}d_i\right)^2 .
$$

(2.50)

Obviously, the left hand side of (2.50) is the time derivative of V along a solution of (2.49) and hence with V as the storage function the dissipativity with respect to (2.48) is shown.   ∎

Remark 2.4. For $d = 0$ we see that the closed-loop (2.49) coincides with (2.13). Hence the origin of (2.49) is asymptotically stable if the zero-state observability of (2.12) from Theorem 2.1 is satisfied.

Again we obtain a pleasing result for linear systems of the type

$$
\frac{d}{dt}x = Ax + Bu + Kd
$$
$$
y = Cx ,
$$

(2.51)

since the special choice $V = \frac{1}{2}x^T P x$, $P > 0$, simplifies the HJBIe (2.45) for $\beta = 1$ to the algebraic Riccati equation

$$
x^T\left(PA + A^T P + C^T C - P\left(BB^T - \frac{1}{\gamma}KK^T\right)P\right)x = 0
$$

(2.52)

or the HJBIe (2.46) to the inequality

$$
x^T\left(PA + A^T P + C^T C - P\left(BB^T - \frac{1}{\gamma}KK^T\right)P\right)x \leq 0 ,
$$

(2.53)

respectively.

Next, we will again investigate systems of the PCHD-type (see (1.73), (1.74) of Chapter 1)

$$
\frac{d}{dt}x = (J(x) - S(x))\left(\frac{\partial V}{\partial x}\right)^T + G_u(x)u + G_d(x)d
$$
$$
y = G_u^T(x)\left(\frac{\partial V}{\partial x}\right)^T .
$$

(2.54)

If the disturbance input d acts in the same way on the system as the control input u, i.e. $G_d(x) = \sqrt{\delta}G_u(x)$, $\delta > 0$, then with the storage function $\rho V(x)$, $\rho > 0$, the HJBIi (2.46) takes the form

$$-\left(\frac{\partial V}{\partial x}\right)\left(\rho S(x) + \frac{1}{2}\left(\beta - \rho^2 + \frac{\rho^2 \delta}{\gamma}\right)G_u(x)G_u^T(x)\right)\left(\frac{\partial V}{\partial x}\right)^T \leq 0.$$

(2.55)

From (2.55) it is clear that the control law

$$u_j = -\rho L_{g_j}V, \; j = 1, \ldots, m$$

(2.56)

solves the suboptimal non-linear H_∞-design problem for a disturbance attenuation level $\gamma > \delta$, provided that the inequality condition

$$\rho \leq \sqrt{\frac{\beta\gamma}{\gamma - \delta}}$$

(2.57)

is satisfied. Moreover, if (2.54) is lossless, i.e. $S(x) = 0$ and (2.57) holds with equality, the storage function $\rho V(x)$ even solves the HJBIe (2.45), see also [143] and for a detailed investigation for infinite-dimensional Hamiltonian systems Chapter 4, in particular, Proposition 4.2.

Remark 2.5. As was already mentioned at the beginning of this chapter, the non-linear H_∞-approach presented here is just what we need in the subsequent chapters for the controller design. For other aspects like the non-linear H_∞-control with dynamic measurement feedback, the reader is referred to e.g., [8], [53], [142], [143], the robust stabilization of perturbed plants, see, e.g., [47], [143] and for the adaptive non-linear H_∞-design, see, e.g., [84].

2.4 Passivity-based Control (PBC)

In recent years, passivity-based control has been, and is still, a field of extensive research and it goes without saying that we can only present a small part of the whole theory. In particular, in this subsection we will refer to that theoretical background material of passivity-based control which will be required in the next chapters. The reader should consult the very important books [107] and [143] and the references cited there for a full overview of this theory and its many interesting applications.

Let us consider a PCHD-system of the form

$$\frac{d}{dt}x = (J(x) - S(x))\left(\frac{\partial V}{\partial x}\right)^T + G_u(x)u$$

(2.58)

with the state $x \in \mathcal{X} \subset R^n$, the control input $u \in \mathcal{U} \subset R^m$, the C^1 positive definite storage function $V(x)$, $V(0) = 0$ and the matrices $G_u(x)$, $J(x) = -J^T(x)$ and $S(x) = S^T(x) \geq 0$ whose entries are smoothly depending on x.

The objective of the PBC-concept, which will be considered here (see [143]), is to find a state feedback $u = u(x)$ such that

- the closed-loop takes the form

$$\frac{d}{dt}x = (J(x) - S(x)) \left(\frac{\partial V_c}{\partial x}\right)^T \tag{2.59}$$

with the modified storage function of the closed-loop $V_c = V + V_a$, where V_a is the contribution of the control input and
- the desired equilibrium $x = x_d$ is stable in the sense of Lyapunov.

Clearly, in order to obtain a closed-loop of the form (2.59) the control law must satisfy the relation

$$G_u(x)u = (J(x) - S(x)) \left(\frac{\partial V_a}{\partial x}\right)^T . \tag{2.60}$$

If $V_c(x)$ has a strict minimum at x_d, that means $V_c(x) > V_c(x_d)$ for all $x \neq x_d$, $\frac{\partial V_c}{\partial x}(x_d) = 0$ and $\frac{\partial^2 V_c}{\partial x^2}(x_d) > 0$, then $V_c(x) - V_c(x_d)$ is positive definite and serves as an appropriate Lyapunov function for the closed-loop (2.59). The time derivative of $V_c(x) - V_c(x_d)$ along an integral curve of (2.59)

$$\frac{d}{dt}V_c = -\left(\frac{\partial V_c}{\partial x}\right) S(x) \left(\frac{\partial V_c}{\partial x}\right)^T \leq 0 \tag{2.61}$$

is obviously negative semi-definite and this guarantees the stability of x_d. Moreover, in some cases we may even show by means of LaSalle's invariance principle that x_d is asymptotically stable.

A further possible way to achieve asymptotic stability is given by the so-called damping injection method [107], [143]. Thereby, the control law u is extended by an additional control input u_d and the relation (2.60) changes to

$$G_u(x)u = (J(x) - S(x)) \left(\frac{\partial V_a}{\partial x}\right)^T + G_u(x)u_d . \tag{2.62}$$

Let us assume that we can measure the output $y = G_u^T(x) \left(\frac{\partial V_c}{\partial x}\right)^T$, see also (1.74) of Chapter 1. Then by substituting (2.62) into (2.58), we obtain the system

$$\frac{d}{dt}x = (J(x) - S(x)) \left(\frac{\partial V_c}{\partial x}\right)^T + G_u(x)u_d$$

$$y = G_u^T(x) \left(\frac{\partial V_c}{\partial x}\right)^T \tag{2.63}$$

with $V_c = V + V_a$. The control law

$$u_d = -S_d\left(x\right) G_u^T\left(x\right) \left(\frac{\partial V_c}{\partial x}\right)^T \tag{2.64}$$

with a suitable positive semi-definite matrix $S_d\left(x\right)$ is commonly referred to as the damping injection controller. The time derivative of V_c along a solution of the closed-loop (2.63) and (2.64) reads as

$$\frac{\mathrm{d}}{\mathrm{d}t}V_c = -\left(\frac{\partial V_c}{\partial x}\right) \underbrace{\left(S\left(x\right) + G_u\left(x\right) S_d\left(x\right) G_u^T\left(x\right)\right)}_{S_c(x)} \left(\frac{\partial V_c}{\partial x}\right)^T. \tag{2.65}$$

Clearly, if $S_c\left(x\right)$ is positive definite, then x_d is asymptotically stable. A necessary condition for the existence of an asymptotically stabilizing control law for the system (2.58) can be found in e.g., [143], Proposition 4.2.14.

Remark 2.6. Apart from this rather simple approach of PBC dynamic controllers with output feedback and controllers for interconnected systems based on the idea of energy-shaping and damping injection are also reported in literature. The latest results in this field are very promising for getting a tool for the design of dynamic non-linear controllers for a large class of physical systems. The interested reader is referred to the very recent publications [108], [109], [143].

3. Electromagnetic Systems

In this chapter, we concentrate on an energy-based description of electrical systems. Thus, an extension of the well-known theory of Brayton-Moser will serve as a basis for setting up the network equations by means of the so-called mixed potentials. It turns out that the combination of this energy-based concept with graph theory allows us to derive the mathematical model of an electrical network directly in the form of a PCHD-system (port-controlled Hamiltonian system with dissipation, see Chapter 1 for details). However, this method is applicable for non-linear two- and three-phase systems with and without dependent sets of inductor currents and/or capacitor voltages. The big advantage of this approach is that it makes available a tool for the systematic derivation of the network equations in the form of a PCHD-system by means of e.g., computer algebra programs. Thus, it is also particularly suitable for larger interconnected networks. Moreover, a very important feature of the proposed concept is that it enables us to take into account saturation effects of the magnetic field and non-sinusoidal flux distributions in three-phase machine applications. The modelling process will be demonstrated by means of a simple terminal model of a power generator and a three-phase application. Furthermore, we will use this technique for the calculation of the average model of PWM (pulse-width-modulation)-controlled electric circuits with bipolar switching, where the duty ratio is the control input. We will see that, depending on the location of the switch(es), different energy flows of the PWM-controlled system can be influenced by changing the duty ratio. Finally, for a laboratory model of a special dc-to-dc converter, namely the Ćuk-converter, we will also show how the presented theory can contribute to the design of a non-linear H_2-controller with and without integral term.

3.1 Basic Circuit Relations

Throughout this section, we consider electric circuits that are interconnections of lumped dynamic (inductors, capacitors) and static (resistors, voltage and current sources) circuit elements. The topological relationship can best be exhibited by means of a digraph $\mathcal{G} = (N, B)$, where N specifies the set of nodes with cardinality n and B denotes the set of branches with cardinality b. A branch has exactly two end points which must be nodes. The orientation of the branches of the digraph corresponds with the associated reference direction of the current flows. Subsequently, we assume that the circuits are connected, i.e. any node can be reached from any other node by a path through the circuit elements. To each branch j a voltage u^j and a current i_j is assigned and due to the connectedness assumption, they are well-defined [19]. Analogously, an electric potential v^k is assigned to each node k. Here, the electrical potential of one arbitrary node has to be chosen as a reference.

Now, Kirchhoff's current law (KCL) says that for lumped circuits the algebraic sum of currents flowing into any node is equal to zero for all times t or

$$\sum_j d_k^j i_j = 0 \quad \text{for all} \quad k \in N. \tag{3.1}$$

Kirchhoff's voltage law (KVL) states that in a lumped circuit for any branch j connected with the nodes k and l the voltage drop u^j is equal to the difference between the potentials v^k and v^l for all times t or

$$v^k d_k^j + v^l d_l^j = u^j \tag{3.2}$$

with

$$d_k^j = \begin{cases} +1 & \text{if branch } j \text{ touches node } k \text{ and } i_j \text{ leaves } k \\ -1 & \text{if branch } j \text{ touches node } k \text{ and } i_j \text{ enters } k \\ 0 & \text{if branch } j \text{ does not touch node } k \,. \end{cases}$$

A current $i^T = (i_1, i_2, \dots, i_b) \in R^b$ and a voltage $u = (u^1, u^2, \dots, u^b) \in (R^b)^*$ with $(R^b)^*$ as the dual space of R^b are said to be admissible if and only if they obey KCL or KVL, respectively. Note that subsequently, in the vector notation, a current is always a column vector and a voltage a row vector. The choice of this notation is motivated by the fact that the voltages are defined in the dual space of the currents or *vice versa*. A direct consequence of Kirchhoff's laws is Tellegen's theorem (e.g., [19], [44]):

Theorem 3.1. *For any admissible current i and any admissible voltage u of an electric circuit with the digraph $\mathcal{G} = (N, B)$ the relation*

$$p = \sum_j u^j i_j = 0 \tag{3.3}$$

holds.

Proof. By substituting (3.2) into (3.3), we get

$$v^k \sum_j d_k^j i_j + v^l \sum_j d_l^j i_j \tag{3.4}$$

and this is obviously zero due to (3.1).   ∎

Remark 3.1. Equation (3.3) is far more than the conservation of energy in an electric circuit. One can easily convince oneself that (3.3) also implies the relations

$$\sum_j u^j (t_1) i_j (t_2) = \sum_j u^j (t_1) \left(\frac{\mathrm{d}}{\mathrm{d}t} i_j \right) (t_2) = \sum_j \left(\frac{\mathrm{d}}{\mathrm{d}t} u^j \right) (t_1) i_j (t_2) = 0$$

(3.5)

for different times t_1 and t_2.

Using the map $D : R^b \to R^n$ with the induced dual map $D^* : (R^n)^* \to (R^b)^*$, we can see that the current i and the voltage u are admissible, if and only if $i \in \mathrm{Ker}\,(D)$ and $u \in \mathrm{Im}\,(D^*)$. It is worth mentioning that, since D is a linear matrix operator, D^* is simply D^T. Let the column vectors of D_I and D_U be the bases of $\mathrm{Ker}\,(D)$ and $\mathrm{Im}\,(D^*)$, respectively. Then the following relations

$$i = D_I i_X \quad \text{and} \quad u = u_Y D_U^T$$

(3.6)

are met for some suitable i_X and u_Y. Combining (3.6) with Tellegen's theorem, we get

$$u D_I = 0 \quad \text{and} \quad D_U^T i = 0 \,.$$

(3.7)

Now, it is well known that D_I and D_U^T can be constructed from a tree $T = (N, B_T)$ with $B_T \subset B$. A tree T of a graph $\mathcal{G}$ is a connected subgraph which has no loops and contains all nodes (see, e.g., [21]). The total number $\rho\,(\mathcal{G})$ of branches in a spanning tree T of the connected graph $\mathcal{G}$, also called co-cyclomatic number, is given by the relation

$$\rho\,(\mathcal{G}) = n - 1$$

(3.8)

with n as the number of nodes. The physical interpretation of the co-cyclomatic number $\rho\,(\mathcal{G})$ in an electrical network is that it gives the largest number of independent potential differences between all the nodes of the network. Analogously, the so-called cyclomatic number

$$\nu\,(\mathcal{G}) = b - \rho\,(\mathcal{G}) = b - n + 1$$

(3.9)

is the largest number of independent currents flowing in the network (see, e.g., [21]). Thus, the graph $\mathcal{G}$ can be partitioned into two disjoint sets of branches, one containing all branches belonging to the tree, the so-called tree branches $B_T \subset B$ with cardinality $\rho\,(\mathcal{G})$ and the other containing all branches which do not belong to the tree, the so-called cotree branches $B_C \subset B$ with cardinality $\nu\,(\mathcal{G})$. For simplification of the notation we will henceforth drop the explicit specification of the associated graph $\mathcal{G}$ in the cyclomatic and co-cyclomatic number ν and ρ. Further, let us combine the voltages and currents of the tree and cotree branches in u_T, u_C and i_T, i_C, respectively.

Consider an electric circuit with associated graph $\mathcal{G}$ and a picked out tree T. If we take $i_X = i_C$ and $u_Y = u_T$ in (3.6), then D_I and D_U^T are well known from literature as the fundamental loop matrix associated with the tree T and the fundamental cut-set matrix associated with the tree T (see, e.g., [19]). Substituting $i = D_I i_C$ and $u = u_T D_U^T$ into (3.7), we obtain

$$u_T D_U^T D_I = 0 \quad \text{and} \quad D_U^T D_I i_C = 0 \tag{3.10}$$

and since (3.10) must be valid for all u_T and i_C, we can deduce

$$D_U^T D_I = 0 . \tag{3.11}$$

If the first ρ branches are the tree branches, then D_I and D_U^T can be partitioned in the form

$$D_I = \begin{bmatrix} D_{I,T} \\ I_{\nu,\nu} \end{bmatrix} \quad , \quad D_U^T = \begin{bmatrix} I_{\rho,\rho}, D_{U,C}^T \end{bmatrix} \tag{3.12}$$

with I as the identity matrix and condition (3.11) can be replaced by

$$D_{I,T} = -D_{U,C}^T = \bar{D} . \tag{3.13}$$

Summarizing (3.6), (3.12) and (3.13), we get

$$i_T = \bar{D} i_C \quad \text{and} \quad u_C = -u_T \bar{D} . \tag{3.14}$$

3.2 Energy Based Description

In this section an energy-based formulation for the state-space equations of electric circuits will be presented. For this purpose, let us consider an electric circuit with associated digraph $\mathcal{G} = (N, B)$. At first, the set of branches B is subdivided into three disjoint sets L, C and S of inductors, capacitors and static terminals, respectively. Clearly, $B = L \cup C \cup S$. According to the notation we use i_k or u^k, $k \in L$, C or S to define the current or voltage of an inductor, capacitor or static terminal, respectively.

The space of unrestricted states (i, u) of the circuit is a manifold $\mathcal{N}_u = R^b \times (R^b)^*$. If the currents and voltages are admissible, or equivalently if they satisfy KCL and KVL, then the states of the electric circuit are confined to a submanifold $\mathcal{N} = \{ (i, u) \in \mathcal{N}_u | u D_I = 0, D_U^T i = 0 \}$ [44].

3.2.1 Independent Set of Inductor Currents and Capacitor Voltages

We assume that the inductor currents i_k, $k \in L$ and the capacitor voltages u^l, $l \in C$ together with the coordinate function $z : (i_k, u^l) \to (i_j, u^j)$, $k \in L$, $l \in C$, $j \in B$ form a chart of $\mathcal{N}$.

The differential equations of the dynamic elements inductor and capacitor are given by

$$\frac{\mathrm{d}}{\mathrm{d}t}\psi^j\left(\ldots,i_k,\ldots\right) = u^j \quad \text{with} \quad j, k \in L$$
$$\frac{\mathrm{d}}{\mathrm{d}t}q_j\left(\ldots,u^k,\ldots\right) = i_j \quad \text{with} \quad j, k \in C,$$

(3.15)

where ψ^j denotes the flux linkage of the inductor $j \in L$ and q_j is the charge linkage of the capacitor $j \in C$. The energies $\hat{w}_L$ and $\hat{w}_C$ stored in the inductors and capacitors can be calculated in the form

$$\hat{w}_L = \int_\gamma \sum_{j \in L} i_j \mathrm{d}\psi^j \quad \text{and} \quad \hat{w}_C = \int_\gamma \sum_{j \in C} u^j \mathrm{d}q_j$$

(3.16)

with $\gamma(t)$ as a solution curve of the electric circuit. The total energy $\hat{w}$ stored in the electric circuit reads as $\hat{w} = \hat{w}_L + \hat{w}_C$. For the subsequent considerations we assume that the integrals in (3.16) are path independent or equivalently the 1-forms $\sum_{j \in L} i_j \mathrm{d}\psi^j$ and $\sum_{j \in C} u^j \mathrm{d}q_j$ are exact. Following Poincare's lemma, which states that in a star-shaped region every closed 1-form is exact, we may deduce that the relations

$$\mathrm{d}\left(\sum_{j \in L} i_j \mathrm{d}\psi^j\right) = 0 \quad \text{and} \quad \mathrm{d}\left(\sum_{j \in C} u^j \mathrm{d}q_j\right) = 0$$

(3.17)

have to be met for the path independence, see, e.g., [16]. Furthermore, the energies $\hat{w}_L$ and $\hat{w}_C$ stored in the inductors and capacitors are always supposed to be positive definite functions of the inductor currents i_k, $k \in L$ and the capacitor voltages u^l, $l \in C$ respectively. Note that this is always the case when the inductor and the capacitor configuration contains magnetic and electrostatic leakage effects, i.e. the inductor and capacitor configurations do not describe ideal transformers.

The static elements in our case resistors, current and voltage sources can be described by the following type of equations

$$f_j\left(\ldots,i_k,\ldots,u^k,\ldots\right) = 0 \quad \text{with} \quad j, k \in S .$$

(3.18)

The flow of power into a terminal $j \in S$ is given by $p_{S,j} = \hat{p}_{S,j} + \check{p}_{S,j}$ with

$$\hat{p}_{S,j} = \int_0^{i_j} u^j \mathrm{d}\tilde{i}_j \quad \text{and} \quad \check{p}_{S,j} = \int_0^{u^j} i_j \mathrm{d}\tilde{u}^j .$$

(3.19)

Also here the assumption of the path independence yields the conditions $\mathrm{d}\left(u^j \mathrm{d}i_j\right) = 0$ and $\mathrm{d}\left(i_j \mathrm{d}u^j\right) = 0$.

Remark 3.2. One can easily conclude that for a voltage source $\check{p}_{S,j} = 0$, for a current source $\hat{p}_{S,j} = 0$ and for a linear resistor $\check{p}_{S,j} = \hat{p}_{S,j}$.

By introducing the expressions

$$p_L = \sum_{j \in L} i_j u^j \quad \text{and} \quad p_C = \sum_{j \in C} i_j u^j, \tag{3.20}$$

we are ready to give a formulation of an extension of Brayton-Moser's well-known theorem which can be found in [120]:

Theorem 3.2. *Given an electric circuit with associated digraph $\mathcal{G} = (N, B)$, $B = L \cup C \cup S$. Let the inductor currents i_k, $k \in L$ and the capacitor voltages u^l, $l \in C$ together with the coordinate function $z : (i_k, u^l) \rightarrow (i_j, u^j)$, $k \in L$, $l \in C$, $j \in B$ be a chart of the configuration manifold $\mathcal{N} = \left\{ (i, u) \in R^b \times (R^b)^* \,\middle|\, u D_I = 0,\, D_U^T i = 0 \right\}$ of the circuit. Let us further suppose that the functions $\hat{w}_L$, $\hat{w}_C$ (3.16) and $\check{p}_S$, $\hat{p}_S$ (3.19) are well defined. Then each trajectory of the electric circuit is a solution curve of*

$$\frac{\mathrm{d}}{\mathrm{d}t} \psi^j = -\frac{\partial}{\partial i_j} (p_C + \hat{p}_S) \quad \text{for} \quad j \in L$$

$$\frac{\mathrm{d}}{\mathrm{d}t} q_j = -\frac{\partial}{\partial u^j} (p_L + \check{p}_S) \quad \text{for} \quad j \in C. \tag{3.21}$$

Furthermore, the relation

$$\frac{\mathrm{d}}{\mathrm{d}t} \hat{w} = -(\check{p}_S + \hat{p}_S) \tag{3.22}$$

is fulfilled for any solution of (3.21). We will call $\hat{p}_S$, $\check{p}_S$, p_C and p_L the mixed potentials of the electric circuit.

Proof. Written in the coordinates (i_k, u^l), $k \in L$, $l \in C$, Tellegen's theorem (see (3.3)) reads as

$$z^* \left(\sum_{j \in B} i_j u^j \right) = 0 \tag{3.23}$$

with z^* as the pullback of $z : (i_k, u^l) \rightarrow (i_j, u^j)$, $k \in L$, $l \in C$, $j \in B$. Of course, (3.23) also implies that

$$z^* \left(\mathrm{d} \sum_{j \in B} i_j u^j \right) = z^* \left(\sum_{j \in B} (i_j \mathrm{d}u^j + u^j \mathrm{d}i_j) \right) = 0 \tag{3.24}$$

and since the forms $\mathrm{d}u^j$ and $\mathrm{d}i_j$ are linearly independent, we have

$$z^* \left(\sum_{j \in B} i_j \mathrm{d}u^j \right) = z^* \left(\sum_{j \in B} u^j \mathrm{d}i_j \right) = 0. \tag{3.25}$$

The first part of (3.25) may be rewritten in the form

$$z^* \left(\sum_{j \in L} i_j du^j + \sum_{j \in C} i_j du^j + \sum_{j \in S} i_j du^j \right) = 0 \tag{3.26}$$

and by applying Leibniz' rule to $\sum_{j \in L}$ and using (3.19) and (3.20), we get

$$z^* \left(d \left(p_L + \check{p}_S \right) - \sum_{j \in L} u^j di_j + \sum_{j \in C} i_j du^j \right) = 0 \tag{3.27}$$

or

$$\sum_{j \in L} z^* \left(\frac{\partial}{\partial i_j} \left(p_L + \check{p}_S \right) - u^j \right) di_j + \sum_{j \in C} z^* \left(\frac{\partial}{\partial u^j} \left(p_L + \check{p}_S \right) + i_j \right) du^j = 0. \tag{3.28}$$

Since the 1-forms di_j, $j \in L$ and du^j, $j \in C$ are linearly independent, the terms in the brackets must vanish identically and hence we get

$$
\begin{aligned}
z^* \left(u^j \right) &= \frac{\partial}{\partial i_j} z^* \left(p_L + \check{p}_S \right) \qquad \text{for} \quad j \in L \\
z^* \left(i_j \right) &= -\frac{\partial}{\partial u^j} z^* \left(p_L + \check{p}_S \right) \quad \text{for} \quad j \in C .
\end{aligned}
\tag{3.29}
$$

Now, the first part (3.21) of the theorem follows directly from the pullback of (3.15)

$$\frac{d}{dt} \psi^j = z^* \left(u^j \right), \quad j \in L \quad \text{and} \quad \frac{d}{dt} q_j = z^* \left(i_j \right), \quad j \in C \tag{3.30}$$

and Tellegen's theorem (3.23), which says that

$$z^* \left(p_L + \check{p}_S \right) = -z^* \left(p_C + \hat{p}_S \right) . \tag{3.31}$$

For the sake of convenience we will subsequently drop the pullback operation z^* in the equations.

Combining (3.16) and (3.31), we get the second relation (3.22) of the theorem in the form

$$\frac{d}{dt} \hat{w} = \frac{d}{dt} \left(\hat{w}_L + \hat{w}_C \right) = p_L + p_C = - \left(\check{p}_S + \hat{p}_S \right) . \tag{3.32}$$

∎

3.2.2 Dependent Set of Inductor Currents and Capacitor Voltages

Theorem 3.2 is only valid if all inductor currents i_j, $j \in L$ and all capacitor voltages u^j, $j \in C$ are linearly independent. Particularly in multi-phase power systems, where we have to deal with Y- and Δ-connected circuits of inductors and capacitors, this is no longer the case. Generally, multi-phase systems are DAEs (differential algebraic equations) but henceforth we will only deal with such systems which allow a description in an explicit form of differential equations. It is worth mentioning that to a certain extent many ideas and concepts, well established in the field of non-linear control for systems of explicit differential equations, can be extended to DAEs, see, e.g., [121], [124]. Subsequently, we will formulate Theorem 3.2 for electric circuits with linear dependent inductors and capacitors. For this purpose we split each of the sets of inductor and capacitor branches $L = L_i \cup L_d$ and $C = C_i \cup C_d$ into a set containing all independent elements L_i, C_i and a set containing all dependent elements L_d and C_d, respectively. Note that in general this task is, of course, not unique. Let us assume that the currents i_k, $k \in L_d$ and the voltages u^m, $m \in C_d$ can be expressed by

$$i_k = \sum_{j \in L_i} \bar{d}_{LL}^{kj} i_j \quad \text{and} \quad u^m = \sum_{l \in C_i} \bar{d}_{CC}^{ml} u^l \tag{3.33}$$

with $\bar{d}_{LL}^{kj}, \bar{d}_{CC}^{ml} \in \{0, 1, -1\}$. Here the inductor currents i_j, $j \in L_i$ and the capacitor voltages u^l, $l \in C_i$ together with the coordinate function $z : (i_k, u^l) \to (i_j, u^j)$, $k \in L_i$, $l \in C_i$, $j \in B$ form a chart of the electric circuit. The differential equations of the dynamic elements, inductor and capacitor, differ slightly from (3.15)

$$\begin{aligned}
\frac{d}{dt} \psi^j \left(\ldots, i_k, \ldots \right) &= u^j \quad \text{with} \quad j \in L, \ k \in L_i \\
\frac{d}{dt} q_j \left(\ldots, u^k, \ldots \right) &= i_j \quad \text{with} \quad j \in C, \ k \in C_i,
\end{aligned} \tag{3.34}$$

because now the flux linkage ψ^j of the inductor $j \in L$ and the charge linkage q_j of the capacitor $j \in C$ depend only on the independent inductor currents i_k, $k \in L_i$ and capacitor voltages u^k, $k \in C_i$. Note that analogous to the previous subsection we also assume here that the energies $\hat{w}_L$ and $\hat{w}_C$ stored in the inductors and capacitors are positive definite functions of i_j, $j \in L_i$ and u^l, $l \in C_i$ respectively. Now, we are able to extend Theorem 3.2 for dependent sets of inductor currents and capacitor voltages.

Theorem 3.3. *Given an electric circuit with associated digraph $\mathcal{G} = (N, B)$, $B = L_i \cup L_d \cup C_i \cup C_d \cup S$. Let the inductor currents i_k, $k \in L_i$ and the capacitor voltages u^l, $l \in C_i$ together with the coordinate function $z : (i_k, u^l) \to (i_j, u^j)$, $k \in L_i$, $l \in C_i$, $j \in B$ be a chart of the configuration manifold $\mathcal{N} = \left\{ (i, u) \in R^b \times \left(R^b \right)^* \,\middle|\, u D_I = 0, \ D_U^T i = 0 \right\}$ of the circuit.*

Let us further suppose that the dependent inductor currents i_k, $k \in L_d$ and capacitor voltages u^l, $l \in C_d$ can be expressed in the form of (3.33) and that the functions $\hat{w}_L$, $\hat{w}_C$ (3.16) and $\check{p}_S$, $\hat{p}_S$ (3.19) are well defined. Then each trajectory of the electric circuit is a solution curve of

$$\frac{d}{dt}\tilde{\psi}^j = -\frac{\partial}{\partial i_j}(p_C + \hat{p}_S) \quad \text{for} \quad j \in L_i$$
$$\frac{d}{dt}\tilde{q}_j = -\frac{\partial}{\partial u^j}(p_L + \check{p}_S) \quad \text{for} \quad j \in C_i \tag{3.35}$$

with $\tilde{\psi}^j = \psi^j + \sum_{k \in L_d} \bar{d}_{LL}^{jk}\psi^k$ and $\tilde{q}_j = q_j + \sum_{m \in C_d} \bar{d}_{CC}^{jm}q_m$, provided that the initial conditions are consistent with the algebraic constraints (3.33). Furthermore, the relation

$$\frac{d}{dt}\hat{w} = -(\check{p}_S + \hat{p}_S) \tag{3.36}$$

is fulfilled for any solution of (3.21). Again, we will call $\hat{p}_S$, $\check{p}_S$, p_C and p_L the mixed potentials of the electric circuit.

Proof. The proof is quite easy and similar to the one of Theorem 3.2. We just have to rewrite (3.27) in the form

$$z^*\left(d\left(p_L + \check{p}_S\right) - \sum_{j \in L_i} u^j di_j - \sum_{k \in L_d} u^k di_k\right) +$$
$$z^*\left(\sum_{l \in C_i} i_l du^l + \sum_{m \in C_d} i_m du^m\right) = 0 \tag{3.37}$$

with z^* as the pullback of $z : (i_k, u^l) \to (i_j, u^j)$, $k \in L_i$, $l \in C_i$, $j \in B$. By substituting relation (3.33) into (3.37), we get

$$z^*\left(d\left(p_L + \check{p}_S\right) - \sum_{j \in L_i}\left(u^j + \sum_{k \in L_d} u^k \bar{d}_{LL}^{kj}\right) di_j\right) +$$
$$z^*\left(\sum_{l \in C_i}\left(i_l + \sum_{m \in C_d} i_m \bar{d}_{CC}^{ml}\right) du^l\right) = 0 \tag{3.38}$$

or

$$\sum_{j \in L_i}\left(\frac{\partial}{\partial i_j}z^*(p_L + \check{p}_S) - z^*\left(u^j + \sum_{k \in L_d} u^k \bar{d}_{LL}^{kj}\right)\right) di_j +$$
$$\sum_{l \in C_i}\left(\frac{\partial}{\partial u^l}z^*(p_L + \check{p}_S) + z^*\left(i_l + \sum_{m \in C_d} i_m \bar{d}_{CC}^{ml}\right)\right) du^l = 0 \,. \tag{3.39}$$

Since the 1-forms di_j, $j \in L_i$ and du^l, $l \in C_i$ are linearly independent, the terms in the brackets must vanish identically. Combining (3.39) with (3.31) and dropping the pullback operation z^*, we obtain directly the result (3.35).

∎

3.3 Energy Based Description with Full Topological Information

3.3.1 Dissipative Electrical Systems in the Form of a PCHD-system

Theorems 3.2 and 3.3 were derived without using the full topological information. In this section, we additionally take advantage of the topology of the electric circuit and this will lead to some pleasing results. Let us consider an electric circuit with associated graph $\mathcal{G}$ and a picked out tree $\mathcal{T}$. It is quite clear that the choice of the tree $\mathcal{T}$ is not unique. Therefore, we are looking for a tree that optimally fits the state-space representation of the electric circuit. Here, we follow the so-called state-tree formulation of [85]. Thereby, the tree is chosen in such a way that the maximum number of capacitors is included in the tree and the maximum number of inductor branches is put in the cotree. The only situation in which all of the capacitors cannot be included in the tree arises from a capacitor-only loop. Similarly, the only reason for an inductor to be placed in the tree is an inductor-only cut-set. If the situation of a capacitor-only loop or of an inductor-only cut-set occurs, then an independent set of capacitors or inductors can always be found for the tree or cotree. The so-called "excess" capacitors or inductors are correspondingly included in the cotree or tree (see, [85] or [94]). Utilizing the state-tree representation, (3.14) can be expanded in the form

$$
\begin{bmatrix} i_{T,C} \\ i_{T,S} \\ i_{T,L} \end{bmatrix} = \begin{bmatrix} \bar{D}_{CL} & \bar{D}_{CS} & \bar{D}_{CC} \\ \bar{D}_{SL} & \bar{D}_{SS} & 0 \\ \bar{D}_{LL} & 0 & 0 \end{bmatrix} \begin{bmatrix} i_{C,L} \\ i_{C,S} \\ i_{C,C} \end{bmatrix} \tag{3.40}
$$

and

$$
[u_{C,L} \; u_{C,S} \; u_{C,C}] = -[u_{T,C} \; u_{T,S} \; u_{T,L}] \begin{bmatrix} \bar{D}_{CL} & \bar{D}_{CS} & \bar{D}_{CC} \\ \bar{D}_{SL} & \bar{D}_{SS} & 0 \\ \bar{D}_{LL} & 0 & 0 \end{bmatrix}, \tag{3.41}
$$

where $i_{T,C}$ and $u_{T,C}$ are the currents and voltages of the tree capacitors; $i_{T,S}$ and $u_{T,S}$ denote the currents and voltages of the static tree elements (resistors, uncontrolled sources); $i_{T,L}$ and $u_{T,L}$ are the currents and voltages of the "excess" tree inductors; $i_{C,L}$ and $u_{C,L}$ are the currents and voltages of the cotree inductors; $i_{C,S}$ and $u_{C,S}$ denote the currents and voltages of the static cotree elements (resistors, uncontrolled sources); and $i_{C,C}$ and $u_{C,C}$ are the currents and voltages of the "excess" cotree capacitors. Furthermore, we assume that all voltage and current sources are placed in the tree and cotree. The only situation in which this assumption cannot be fulfilled is a loop consisting entirely of capacitors and voltage sources, or a cut-set consisting entirely of inductors and current sources. In such a situation a unit step function generated by the voltage or current source would cause an unbounded capacitor current or inductor voltage, respectively. Hence, this assumption is no restriction of generality at all. For the static tree and cotree elements two further assumptions are made, namely

- the resistors are linear and
- only uncontrolled voltage and current sources are allowed.

Under these conditions we may partition the voltages of the static tree elements $u_{T,S}$ and the currents of the static cotree elements $i_{C,S}$ in such a way that they possess the following functional form

$$u_{T,S}^T = \underbrace{\begin{bmatrix} \bar{R} & 0 \\ 0 & 0 \end{bmatrix}}_{R} \dot{i}_{T,S} + \begin{bmatrix} 0 \\ U_0^T \end{bmatrix}, \quad i_{C,S} = \underbrace{\begin{bmatrix} \bar{G} & 0 \\ 0 & 0 \end{bmatrix}}_{G} u_{C,S}^T + \begin{bmatrix} 0 \\ I_0 \end{bmatrix}. \tag{3.42}$$

Here U_0 and I_0 combine the voltages and currents of the uncontrolled voltage and current sources and $\bar{R}$ and $\bar{G}$ are positive semi-definite diagonal matrices containing the resistors and conductances of the tree and cotree, respectively. At this point it should be emphasized that in general an electric network consisting of arbitrary non-linear static and dynamic elements is described by DAEs (differential algebraic equations) and cannot be transformed into an explicit form. Note again that in our notation, currents and charges are always arranged in column vectors and voltages and fluxes in row vectors. For electric circuits satisfying the above properties, we are able to formulate the following proposition:

Proposition 3.1. *Given an electric circuit with digraph $\mathcal{G} = (N, B)$, $B = L \cup C \cup S$, let each trajectory of the electric circuit be a solution curve of (3.35) and let all initial conditions be consistent with the algebraic constraints (3.33). Let us further suppose, we have picked out a tree T of $\mathcal{G}$ with the state-tree representation of (3.40) and (3.41) and the voltages and currents of the static tree and cotree elements possess the functional form of (3.42). Then the network equations can be directly expressed in the form of a PCHD-system (see (1.73) of Chapter 1)*

$$\frac{d}{dt} x = (J - S) \left(\frac{\partial \hat{w}}{\partial x} \right)^T + G_{U_0} \begin{bmatrix} 0 \\ U_0^T \end{bmatrix} + G_{I_0} \begin{bmatrix} 0 \\ I_0 \end{bmatrix} \tag{3.43}$$

with the state $x^T = \begin{bmatrix} \tilde{\psi}_{C,L}, \tilde{q}_{T,C}^T \end{bmatrix}$, $\tilde{\psi}_{C,L} = \psi_{C,L} + \psi_{T,L} \bar{D}_{LL}$, $\tilde{q}_{T,C} = q_{T,C} - \bar{D}_{CC} q_{C,C}$, *the total energy* $\hat{w} = \hat{w}_L + \hat{w}_C$ *stored in the inductors and capacitors and*

$$J = \begin{bmatrix} 0 & \Lambda \\ -\Lambda^T & 0 \end{bmatrix}, \quad S = \begin{bmatrix} S_1 & 0 \\ 0 & S_2 \end{bmatrix}, \tag{3.44}$$

$$G_{U_0} = \begin{bmatrix} -\bar{D}_{SL}^T \Pi_1 \\ -\bar{D}_{CS}\Pi_2 G\bar{D}_{SS}^T \end{bmatrix}, \quad G_{I_0} = \begin{bmatrix} -\bar{D}_{SL}^T \Pi_1 R\bar{D}_{SS} \\ \bar{D}_{CS}\Pi_2 \end{bmatrix}, \tag{3.45}$$

where

$$\begin{aligned}
\Lambda &= (-\bar{D}_{CL}^T + \bar{D}_{SL}^T R\bar{D}_{SS}G\Pi_2^T \bar{D}_{CS}^T) \\
S_1 &= \bar{D}_{SL}^T \Pi_1 R\bar{D}_{SL} \\
S_2 &= \bar{D}_{CS}\Pi_2 G\bar{D}_{CS}^T \\
\Pi_1 &= (I + R\bar{D}_{SS}G\bar{D}_{SS}^T)^{-1} \\
\Pi_2 &= (I + G\bar{D}_{SS}^T R\bar{D}_{SS})^{-1}
\end{aligned} \tag{3.46}$$

and

$$R = \begin{bmatrix} \bar{R} & 0 \\ 0 & 0 \end{bmatrix}, \quad G = \begin{bmatrix} \bar{G} & 0 \\ 0 & 0 \end{bmatrix}. \tag{3.47}$$

Obviously, the matrices S_1, S_2, R and G are positive semi-definite.

Proof. The basis of the proof is Theorem 3.3. Here, the independent inductor currents i_k, $k \in L_i$ are combined in $i_{C,L}$ and the independent capacitor voltages u^l, $l \in C_i$ in $u_{T,C}$. Using the state-tree representation (3.40) and (3.41), p_C and p_L from (3.20) take the form

$$\begin{aligned}
p_C &= u_{T,C}i_{T,C} + u_{C,C}i_{C,C} = u_{T,C}\left(\bar{D}_{CL}i_{C,L} + \bar{D}_{CS}i_{C,S}\right) \\
p_L &= u_{C,L}i_{C,L} + u_{T,L}i_{T,L} = -\left(u_{T,C}\bar{D}_{CL} + u_{T,S}\bar{D}_{SL}\right)i_{C,L}.
\end{aligned} \tag{3.48}$$

The expressions $\hat{p}_S$ and $\check{p}_S$ can be written as

$$\hat{p}_S = \hat{p}_S^{U_0} + \hat{p}_S^R \quad \text{and} \quad \check{p}_S = \check{p}_S^{I_0} + \check{p}_S^R, \tag{3.49}$$

where $\hat{p}_S^{U_0}$ and $\check{p}_S^{I_0}$ are the parts due to the independent voltage and current sources and $\hat{p}_S^R$, $\check{p}_S^R$ represent the resistors of the tree and cotree, respectively. By (3.42) we may deduce

$$\hat{p}_S^R = \check{p}_S^R = \frac{1}{2}i_{T,S}^T R i_{T,S} + \frac{1}{2}u_{C,S}Gu_{C,S}^T. \tag{3.50}$$

Now, in a first step we calculate $\dfrac{\partial p_C}{\partial i_{C,L}}$ and $\dfrac{\partial p_L}{\partial u_{T,C}}$. Using (3.48) in combination with (3.40) and (3.41), we get

$$\frac{\partial p_C}{\partial i_{C,L}} = u_{T,C}\left(\bar{D}_{CL} + \bar{D}_{CS}\frac{\partial i_{C,S}}{\partial i_{C,L}}\right) \tag{3.51}$$

with

$$\frac{\partial i_{C,S}}{\partial i_{C,L}} = -G\bar{D}_{SS}^T \frac{\partial u_{T,S}^T}{\partial i_{C,L}} = -G\bar{D}_{SS}^T R\left(\bar{D}_{SL} + \bar{D}_{SS}\frac{\partial i_{C,S}}{\partial i_{C,L}}\right) \tag{3.52}$$

or

$$\frac{\partial i_{C,S}}{\partial i_{C,L}} = -\left(I + G\bar{D}_{SS}^T R \bar{D}_{SS}\right)^{-1} G\bar{D}_{SS}^T R \bar{D}_{SL} \tag{3.53}$$

and hence

$$\left(\frac{\partial p_C}{\partial i_{C,L}}\right)^T = A u_{T,C}^T \tag{3.54}$$

with

$$A = \left(\bar{D}_{CL}^T - \bar{D}_{SL}^T R \bar{D}_{SS} G \left(I + \bar{D}_{SS}^T R \bar{D}_{SS} G\right)^{-1} \bar{D}_{CS}^T\right) \tag{3.55}$$

and

$$R = \begin{bmatrix} \bar{R} & 0 \\ 0 & 0 \end{bmatrix} \quad \text{and} \quad G = \begin{bmatrix} \bar{G} & 0 \\ 0 & 0 \end{bmatrix} . \tag{3.56}$$

In this way, we also get

$$\frac{\partial p_L}{\partial u_{T,C}} = -i_{C,L}^T \left(\bar{D}_{CL}^T + \bar{D}_{SL}^T \frac{\partial u_{T,S}^T}{\partial u_{T,C}}\right) \tag{3.57}$$

with

$$\frac{\partial u_{T,S}^T}{\partial u_{T,C}} = -\left(I + R\bar{D}_{SS} G\bar{D}_{SS}^T\right)^{-1} R\bar{D}_{SS} G\bar{D}_{CS}^T \tag{3.58}$$

and hence

$$\left(\frac{\partial p_L}{\partial u_{T,C}}\right)^T = B i_{C,L}, \tag{3.59}$$

where

$$B = \left(-\bar{D}_{CL} + \bar{D}_{CS} G\bar{D}_{SS}^T R \left(I + \bar{D}_{SS} G\bar{D}_{SS}^T R\right)^{-1} \bar{D}_{SL}\right) . \tag{3.60}$$

Next, we show that $A = -\Lambda$ and $B = \Lambda^T$. Obviously, this is the fact if

$$\left(I + G\bar{D}_{SS}^T R \bar{D}_{SS}\right)^{-1} G\bar{D}_{SS}^T R = G\bar{D}_{SS}^T R \left(I + \bar{D}_{SS} G\bar{D}_{SS}^T R\right)^{-1} . \tag{3.61}$$

Multiplying with $\left(I + G\bar{D}_{SS}^T R \bar{D}_{SS}\right)$ from the left and from the right with $\left(I + \bar{D}_{SS} G\bar{D}_{SS}^T R\right)$, we directly obtain the identity

$$G\bar{D}_{SS}^T R \left(I + \bar{D}_{SS} G\bar{D}_{SS}^T R\right) = \left(I + G\bar{D}_{SS}^T R \bar{D}_{SS}\right) G\bar{D}_{SS}^T R . \tag{3.62}$$

Up to now, the energy preserving part of the system was derived. In a second step, we shall determine $\dfrac{\partial \hat{p}_S}{\partial i_{C,L}}$ and $\dfrac{\partial \check{p}_S}{\partial u_{T,C}}$. From (3.49), it follows

$$\frac{\partial \hat{p}_S}{\partial i_{C,L}} = \frac{\partial \hat{p}_S^{U_0}}{\partial i_{C,L}} + \frac{\partial \hat{p}_S^R}{\partial i_{C,L}} \quad \text{and} \quad \frac{\partial \check{p}_S}{\partial u_{T,C}} = \frac{\partial \check{p}_S^{I_0}}{\partial u_{T,C}} + \frac{\partial \check{p}_S^R}{\partial u_{T,C}} \tag{3.63}$$

and by using (3.50) and (3.42), we get

$$\left(\frac{\partial \hat{p}_S^{U_0}}{\partial i_{C,L}}\right)^T = \left(\frac{\partial i_{T,S}}{\partial i_{C,L}}\right)^T \begin{bmatrix} 0 \\ U_0^T \end{bmatrix} = \bar{D}_{SL}^T \left(I + R\bar{D}_{SS}G\bar{D}_{SS}^T\right)^{-1} \begin{bmatrix} 0 \\ U_0^T \end{bmatrix} \tag{3.64}$$

and

$$\left(\frac{\partial \check{p}_S^{I_0}}{\partial u_{T,C}}\right)^T = \left(\frac{\partial u_{C,S}^T}{\partial u_{T,C}}\right)^T \begin{bmatrix} 0 \\ I_0 \end{bmatrix} = -\bar{D}_{CS} \left(I + G\bar{D}_{SS}^T R\bar{D}_{SS}\right)^{-1} \begin{bmatrix} 0 \\ I_0 \end{bmatrix} . \tag{3.65}$$

With (3.50) the second part of (3.63) reads as

$$\frac{\partial \hat{p}_S^R}{\partial i_{C,L}} = i_{T,S}^T R \frac{\partial i_{T,S}}{\partial i_{C,L}} + u_{C,S} G \frac{\partial u_{C,S}^T}{\partial i_{C,L}} \tag{3.66}$$

or after some simplifications due to (3.40) and (3.41)

$$\frac{\partial \hat{p}_S^R}{\partial i_{C,L}} = \left([0, \ I_0^T] \ \bar{D}_{SS}^T R + i_{C,L}^T \bar{D}_{SL}^T R\right) \frac{\partial i_{T,S}}{\partial i_{C,L}}. \tag{3.67}$$

Finally, we have

$$\left(\frac{\partial \hat{p}_S^R}{\partial i_{C,L}}\right)^T = \bar{D}_{SL}^T \left(I + R\bar{D}_{SS}G\bar{D}_{SS}^T\right)^{-1} \left(R\bar{D}_{SL}i_{C,L} + R\bar{D}_{SS} \begin{bmatrix} 0 \\ I_0 \end{bmatrix}\right) . \tag{3.68}$$

Analogously, with

$$\frac{\partial \check{p}_S^R}{\partial u_{T,C}} = i_{T,S}^T R \frac{\partial i_{T,S}}{\partial u_{T,C}} + u_{C,S} G \frac{\partial u_{C,S}^T}{\partial u_{T,C}} \tag{3.69}$$

or

$$\frac{\partial \check{p}_S^R}{\partial u_{T,C}} = -\left(u_{T,C}\bar{D}_{CS}G + [0, \ U_0] \ \bar{D}_{SS}^T G\right) \frac{\partial u_{C,S}^T}{\partial u_{T,C}} \tag{3.70}$$

the result reads as

$$\left(\frac{\partial \check{p}_S^R}{\partial u_{T,C}}\right)^T = \bar{D}_{CS} \left(I + G\bar{D}_{SS}^T R\bar{D}_{SS}\right)^{-1} \left(G\bar{D}_{CS}^T u_{T,C}^T + G\bar{D}_{SS}^T \begin{bmatrix} 0 \\ U_0^T \end{bmatrix}\right) . \tag{3.71}$$

Substituting (3.54), (3.59), (3.64), (3.65), (3.68) and (3.71) into (3.35), we immediately get (3.43). The only thing that remains to be shown is $\left(\frac{\partial \hat{w}}{\partial x}\right) = \left[i_{C,L}^{T}, u_{T,C}\right]$ with $x^{T} = \left[\tilde{\psi}_{C,L}, \tilde{q}_{T,C}^{T}\right]$, $\tilde{\psi}_{C,L} = \psi_{C,L} + \psi_{T,L}\bar{D}_{LL}$, $\tilde{q}_{T,C} = q_{T,C} - \bar{D}_{CC}q_{C,C}$. By means of the state-tree representation of (3.40) and (3.41) we get from (3.16) the energy stored in the inductors

$$
\begin{aligned}
\hat{w}_L &= \int_\gamma \left(i_{C,L}^T \mathrm{d}\psi_{C,L}^T + i_{T,L}^T \mathrm{d}\psi_{T,L}^T \right) = \int_\gamma i_{C,L}^T \mathrm{d}\left(\psi_{C,L}^T + \bar{D}_{LL}^T \psi_{T,L}^T \right) \\
&= \int_\gamma i_{C,L}^T \mathrm{d}\tilde{\psi}_{C,L}^T
\end{aligned}
$$
(3.72)

and in the capacitors

$$
\begin{aligned}
\hat{w}_C &= \int_\gamma \left(u_{T,C} \mathrm{d}q_{T,C} + u_{C,C} \mathrm{d}q_{C,C} \right) = \int_\gamma u_{T,C} \mathrm{d}\left(q_{T,C} - \bar{D}_{CC}q_{C,C} \right) \\
&= \int_\gamma u_{T,C} \mathrm{d}\tilde{q}_{T,C} \ .
\end{aligned}
$$
(3.73)

From (3.72) and (3.73), we can conclude that

$$
\frac{\partial \hat{w}_L}{\partial \tilde{\psi}_{C,L}^T} = i_{C,L}^T \quad \text{and} \quad \frac{\partial \hat{w}_C}{\partial \tilde{q}_{T,C}} = u_{T,C} \tag{3.74}
$$

and this completes the proof.   ∎

It is worth mentioning that by means of suitable algorithms from graph theory, like the search method depth-first (see, e.g., [21]), the state-tree representation of (3.40) and (3.41) can be easily obtained in a systematic way. Within the scope of a controller design, not only the numeric simulation of the electric circuit, but also the symbolic calculation of the circuit equations is of importance. The proposed methods are especially useful for computer algebra implementation, and in combination with object oriented programming even electrical networks of up to fifty branches are symbolically calculable [61].

3.3.2 Application: Simple Electric Circuit

Let us at first demonstrate this procedure by means of the simple electric circuit of Fig. 3.1 which represents a simple terminal model for a power generator with mutual phase coupling and a voltage applied between one phase and the ground [37]. Fig. 3.2 depicts the associated digraph $\mathcal{G} = (N, B)$ and the tree $\mathcal{T}$, indicated by the dotted lines, for the state-tree representation. The orientation of the branches corresponds with the associated reference direction of the current flows. Utilizing the state-tree of Fig. 3.2, we can write (3.40) and R, G and U_0^T from (3.42) for this simple electric circuit in the form

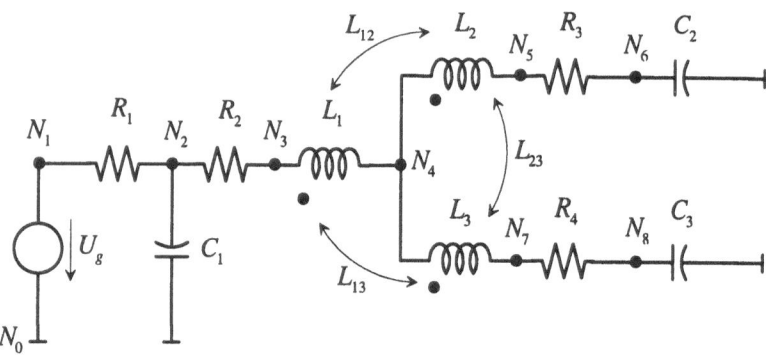

Fig. 3.1. Simple electric circuit.

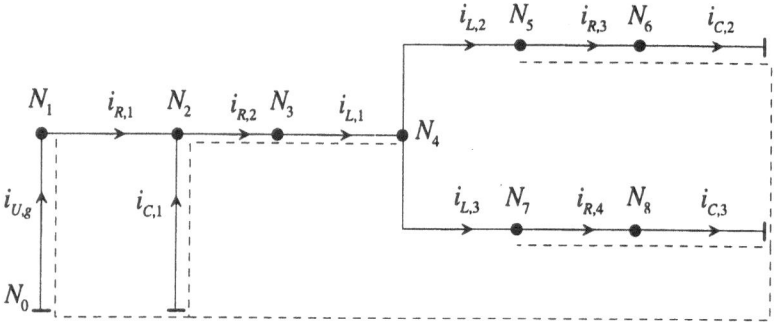

Fig. 3.2. Digraph and state-tree for Fig. 3.1.

$$
\begin{bmatrix} i_{C,1} \\ i_{C,2} \\ i_{C,3} \\ i_{R,2} \\ i_{R,3} \\ i_{R,4} \\ i_{U,g} \\ i_{L,1} \end{bmatrix} = \begin{bmatrix} 1 & 1 & -1 \\ 1 & 0 & 0 \\ 0 & 1 & 0 \\ 1 & 1 & 0 \\ 1 & 0 & 0 \\ 0 & 1 & 0 \\ 0 & 0 & 1 \\ 1 & 1 & 0 \end{bmatrix} \begin{bmatrix} i_{L,2} \\ i_{L,3} \\ i_{R,1} \end{bmatrix} \quad \text{and} \quad
\begin{aligned}
R &= \begin{bmatrix} R_2 & 0 & 0 & 0 \\ 0 & R_3 & 0 & 0 \\ 0 & 0 & R_4 & 0 \\ 0 & 0 & 0 & 0 \end{bmatrix} \\[2mm]
G &= \begin{bmatrix} \dfrac{1}{R_1} \end{bmatrix}, \; U_0^T = [-U_g] \; .
\end{aligned}
\tag{3.75}
$$

The matrices $\bar{D}_{CL}$, $\bar{D}_{CS}$, $\bar{D}_{SL}$, $\bar{D}_{SS}$ and $\bar{D}_{LL}$ can be directly obtained by a suitable partitioning of (3.75). Applying Proposition 3.1, we get the circuit equations

$$
\frac{\mathrm{d}}{\mathrm{dt}} x = \left(\begin{bmatrix} 0 & \Lambda \\ -\Lambda^T & 0 \end{bmatrix} - \begin{bmatrix} S_1 & 0 \\ 0 & S_2 \end{bmatrix} \right) \left(\frac{\partial \hat{w}}{\partial x} \right)^T + \begin{bmatrix} 0 \\ G_{e,C} \end{bmatrix} U_g
\tag{3.76}
$$

with the state $x^T = \begin{bmatrix} \tilde{\psi}_{L,2}, & \tilde{\psi}_{L,3}, & q_{C,1}, & q_{C,2}, & q_{C,3} \end{bmatrix}$,

$$\tilde{\psi}_{L,2} = (L_1 + L_2 + 2L_{12})\, i_{L,2} + (L_1 + L_{12} + L_{13} + L_{23})\, i_{L,3}$$
$$\tilde{\psi}_{L,3} = (L_1 + L_{12} + L_{13} + L_{23})\, i_{L,2} + (L_1 + L_3 + 2L_{13})\, i_{L,3} \qquad (3.77)$$
$$q_{C,i} = C_i u_{C,i}, \quad i = 1, 2, 3\ ,$$

and the total stored energy

$$\hat{w} = \frac{(L_1 + L_2 + 2L_{12})}{2} i_{L,2}^2 + \frac{(L_1 + L_3 + 2L_{13})}{2} i_{L,3}^2 + \sum_{i=1}^{3} \frac{C_i}{2} u_{C,i}^2$$
$$+ (L_1 + L_{12} + L_{13} + L_{23})\, i_{L,2} i_{L,3}\ .$$

$$(3.78)$$

The matrices Λ, S_1, S_2 and $G_{e,C}$ are given by

$$\Lambda = \begin{bmatrix} -1 & -1 & 0 \\ -1 & 0 & -1 \end{bmatrix}, \quad S_1 = \begin{bmatrix} R_2 + R_3 & R_2 \\ R_2 & R_2 + R_4 \end{bmatrix}, \qquad (3.79)$$

$$S_2 = \begin{bmatrix} \dfrac{1}{R_1} & 0 & 0 \\ 0 & 0 & 0 \\ 0 & 0 & 0 \end{bmatrix} \quad \text{and} \quad G_{e,C} = \begin{bmatrix} \dfrac{-1}{R_1} \\ 0 \\ 0 \end{bmatrix}. \qquad (3.80)$$

3.3.3 Application: Three-phase Power System

For the steady-state analysis of balanced and unbalanced three-phase power systems, the method of symmetrical components is the first choice (see, e.g., [11]). However, if we are interested in a transient description of three-phase power systems then the use of the symmetrical components is no longer valid. In this case, the energy-based description of electrical systems introduced so far seems to be an appropriate method for the derivation of the mathematical model, in particular for control purposes. The energy-based description enables a good insight into the energy situation of the system and hence also supports the understanding of the system dynamics. Due to its systematic nature the proposed graph-theoretic approach can also be applied to large-scale power transmission and multi-machine systems. Moreover, the resulting energy-based formulation may serve as a basis for the stability analysis of power systems. A very important feature of the proposed concept is that we are able to take into account saturation effects of the magnetic field and non-sinusoidal flux distributions. It is worth mentioning that the transient models of three-phase machine applications with linear magnetic characteristics are often based on the so-called Blondel-Park transformation. Here the mathematical model is transformed into a simple form such that the rotor position dependence disappears in the equations. See [83], under which necessary and sufficient conditions this transformation is possible. In general, these conditions are not satisfied in the case of a non-sinusoidal flux distribution or if the magnetic characteristic is non-linear. However, here the energy-based description discussed so far offers new ways for designing energy-based non-linear controllers which can capture also these effects. Next, we want to give a simple example of such a three-phase power system and its associated state-tree. Fig. 3.3 represents the electric circuit of a stand-alone self-excited induction generator with Δ-connected excitation capacitors and a balanced load consisting of inductors and resistors. Self-excited induction generators with an excitation control can be often found in combination with wind turbines because they are able to generate power at constant voltage and frequency even with varying wind speed and changing load conditions (see, e.g., [114]).

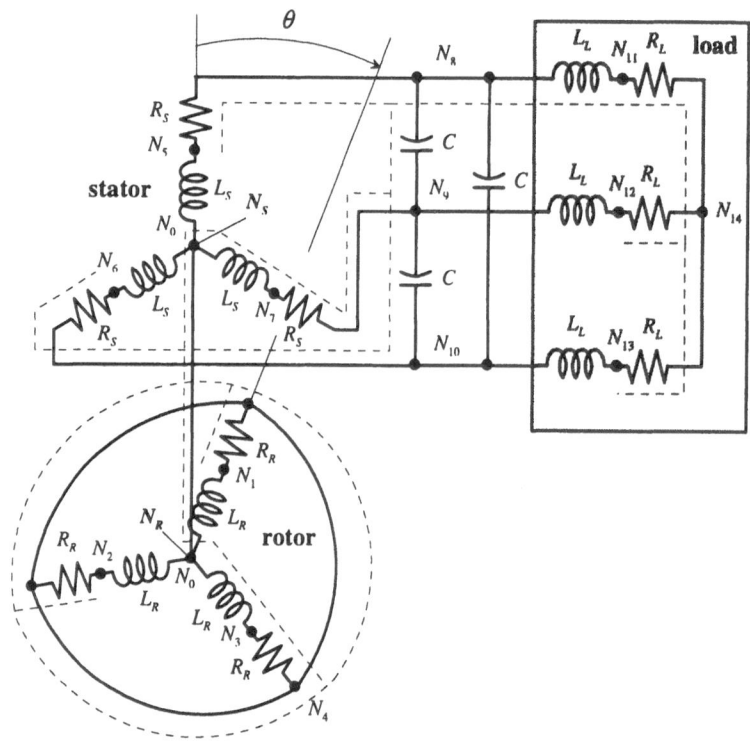

Fig. 3.3. Electric circuit of a stand alone self-excited induction generator with a balanced load.

The stator and rotor of the induction generator are magnetically coupled via the flux linkages. Therefore, we will *a priori* get an unconnected associated digraph which consists of two components, one for the stator and one for the rotor. But without changing the circuit behavior in our case, we are allowed to tie together the neutrals $N_S = N_R = N_0$ of the stator and rotor, respectively. The resulting connected graph is called a hinged graph (see, e.g.,

[19]). As shown in Fig. 3.3 the hinged digraph $\mathcal{G} = (N, B)$ has $n = 15$ nodes, $b = 21$ branches and due to (3.9) the cyclomatic number $\nu(\mathcal{G}) = 7$. The dotted line in Fig. 3.3 marks a possible tree for the state-tree representation (3.40) and (3.41). As one can see, we have three access inductors (inductors in the tree) and one excess capacitor (capacitor in the cotree). Before applying Proposition 3.1, we have to define the flux and charge linkages of the inductors and capacitors respectively. The load inductors and the self-excitation capacitors are assumed not to be mutually coupled. Note that the only restriction of the functional dependence of the stator and rotor flux linkages $\psi_S = [\psi_{S,1}, \psi_{S,2}, \psi_{S,3}]$ and $\psi_R = [\psi_{R,1}, \psi_{R,2}, \psi_{R,3}]$ on the stator and rotor currents $i_S^T = [i_{S,1}, i_{S,2}, i_{S,3}]$ and $i_R^T = [i_{R,1}, i_{R,2}, i_{R,3}]$ and on the angle of rotation of the rotor θ (measured in the stator frame of coordinates) is given by (3.17), i.e.

$$\frac{\partial \psi_{S,j}}{\partial i_{S,i}} = \frac{\partial \psi_{S,i}}{\partial i_{S,j}}, \frac{\partial \psi_{R,j}}{\partial i_{R,i}} = \frac{\partial \psi_{R,i}}{\partial i_{R,j}} \quad \text{and} \quad \frac{\partial \psi_{R,j}}{\partial i_{S,i}} = \frac{\partial \psi_{S,i}}{\partial i_{R,j}} \tag{3.81}$$

for $i \neq j = 1, 2, 3$. Note that the exterior derivative d in (3.17) only operates in the variables (i_k, u^l), $k \in L_i$, $l \in C_i$ and has no effect on θ. Thus, with the proposed method we are even able to take into account saturation of the magnetic field and non-sinusoidal flux distributions. We do not intend to derive the mathematical model in detail here because with the state-tree shown in Fig. 3.3 (dotted line) and Proposition 3.1 it is a straightforward task.

It is worth mentioning that normally for control purposes the induction generator (motor) is considered to be magnetic linear and therefore, the stator and rotor flux linkages can be expressed as

$$\begin{aligned} \psi_S &= L_{SS} i_S + L_{SR}(\theta) i_R \\ \psi_R &= L_{RS}(\theta) i_S + L_{RR} i_R \end{aligned} \tag{3.82}$$

with the symmetric self inductance matrices L_{SS} and L_{RR} of the stator and rotor respectively and the stator-to-rotor mutual inductance matrix $L_{SR}(\theta) = L_{RS}^T(\theta)$. If furthermore, the induction generator has a symmetrical three phase winding and the mutual stator-to-rotor inductance is a harmonic function of the angle θ, then L_{SS}, L_{RR} and $L_{SR}(\theta)$ take the simple form

$$L_{KK} = \begin{bmatrix} L + L_{\sigma K} & -\dfrac{L}{2} & -\dfrac{L}{2} \\ -\dfrac{L}{2} & L + L_{\sigma K} & -\dfrac{L}{2} \\ -\dfrac{L}{2} & -\dfrac{L}{2} & L + L_{\sigma K} \end{bmatrix} \quad \text{with } K \in \{S, R\} \tag{3.83}$$

and

$$L_{SR}(\theta) = \begin{bmatrix} L\cos(\theta) & L\cos(\theta + \gamma) & L\cos(\theta - \gamma) \\ L\cos(\theta - \gamma) & L\cos(\theta) & L\cos(\theta + \gamma) \\ L\cos(\theta + \gamma) & L\cos(\theta - \gamma) & L\cos(\theta) \end{bmatrix} \qquad (3.84)$$

with $\gamma = \frac{2\pi}{3}$, the main inductance L and the leakage inductances of the stator and rotor $L_{\sigma S}$ and $L_{\sigma R}$. In the case when these symmetry conditions are satisfied, the Blondel-Park transformation can be performed (see, e.g., [83]). However, these simplifications, although they are not an essential restriction for many applications, are not necessary for the applicability of the energy-based formulation of Proposition 3.1.

3.4 PWM-controlled Electrical Systems

3.4.1 Energy Based Description

Let us consider a PWM (pulse-width-modulation)-controlled electric circuit with s switches $\{T_1, \ldots, T_s\}$ and each switch T_i, $i = 1, \ldots, s$ has two positions: an on-position denoted by A and an off-position denoted by B. It is assumed that the s-switch tuple $\{T_1, \ldots, T_s\}$ is turned on and off simultaneously. For example, in the case of a full-bridge dc-to-dc converter (see, e.g., [57], [99]) the four switches are treated as one switch tuple, which is either in position A or B. In this case the PWM strategy is also called PWM with bipolar voltage switching [99].

Suppose that the associated digraph of the PWM-controlled circuit is connected for both switch positions, A and B. Let us further assume throughout the whole section that the same inductor currents i_k, $k \in L_i$ and capacitor voltages u^l, $l \in C_i$ together with the coordinate function $z : (i_j, u^j) \to (i_k, u^l)$, $j \in B$, $k \in L_i$, $l \in C_i$ form a chart of the PWM-controlled network in both switch positions, A and B. Moreover, the functional dependence of the dependent inductor currents i_k, $k \in L_d$ and capacitor voltages u^m, $m \in C_d$ due to (3.33) is supposed not to change with the switch position. One can easily convince oneself using practical examples, that these assumptions are not very restrictive. Then, the considered PWM-controlled circuits can be described by two systems of differential equations in the form

$$\frac{d}{dt}x = f_A(x) \quad t \in (iT, (i + d_A)T] \qquad \{T_1, \ldots, T_s\} \text{ in } A$$
$$\frac{d}{dt}x = f_B(x) \quad t \in ((i + d_A)T, (i + d_A + d_B)T] \quad \{T_1, \ldots, T_s\} \text{ in } B$$

$$(3.85)$$

for $i = 0, 1, \ldots$ with the smooth vector fields f_A, f_B and $d_A + d_B = 1$. Here, d_A, $0 \le d_A \le 1$ denotes the so-called duty ratio, which specifies the ratio of the duration of the s-switch tuple $\{T_1, \ldots, T_s\}$ in position A to the fixed modulation period T. Now we know that the state x of (3.85) is continuous because the theory of differential equations says that the state x of a system $\frac{d}{dt}x = f(x, u)$ with piecewise continuous inputs u are continuous [44], [144].

Under the assumption that the modulation frequency is much higher than the natural frequencies of the converter system and the switches are realized with common power semiconductor devices, we can derive the so-called average model for the PWM-controlled electric circuit (3.85), see, e.g., [57], [127]. Let $\varphi_t^{f_A}(x)$ denote the flow of the electric circuit for the s-switch tuple $\{T_1, \dots, T_s\}$ in position A and $\varphi_t^{f_B}(x)$ in position B. Then a solution $\gamma(t)$ of the PWM-controlled electric circuit for $t = iT$, $i = 0, 1, \dots$ meets the relation

$$\gamma((i+1)T) = \varphi_{d_A T}^{f_A} \circ \varphi_{d_B T}^{f_B}(\gamma(iT)) \ , \quad d_A + d_B = 1 \ . \tag{3.86}$$

The average model

$$\frac{d}{dt}x_a = (d_A f_A + d_B f_B)(x_a) \tag{3.87}$$

is nothing else than the first order approximation of $\gamma(t)$ by $\gamma_a(t)$ [116]

$$\frac{d}{dt}\gamma_a(iT) = \lim_{T \to 0} \partial_T \varphi_{d_A T}^{f_A} \circ \varphi_{d_B T}^{f_B}(\gamma_a(iT)) \quad t = iT, \ i = 0, 1, \dots. \tag{3.88}$$

Remark 3.3. By bending the rules of notation we will use the same symbol x for the state variables of (3.85) and for the approximated average model (3.87). Thus, we will subsequently drop the index a.

The energy-based description of Theorem 3.2 and 3.3 can also be applied to derive the average model of a PWM-controlled electric circuit in a straightforward manner. Since it is assumed that $\tilde{\psi}^j$, $j \in L_i$ and $\tilde{q}_j$, $j \in C_i$ from (3.35) do not depend on the switch position, we may write the average model in the form

$$\begin{aligned}
\frac{d}{dt}\tilde{\psi}^j &= -\frac{\partial}{\partial i_j} \sum_{K \in \{A,B\}} (p_C^K + \hat{p}_S^K) d_K \quad \text{for} \quad j \in L_i \\
\frac{d}{dt}\tilde{q}_j &= -\frac{\partial}{\partial u^j} \sum_{K \in \{A,B\}} (p_L^K + \check{p}_S^K) d_K \quad \text{for} \quad j \in C_i
\end{aligned} \tag{3.89}$$

with $d_A + d_B = 1$. Note that $K = A$ or $K = B$ for p_C^K, $\hat{p}_S^K$, p_L^K and $\check{p}_S^K$ refer to the corresponding quantity for the s-switch tuple $\{T_1, \dots, T_s\}$ in position A or B respectively.

3.4.2 Energy Based Description with Full Topological Information

Consider a PWM-controlled circuit with bipolar voltage switching as discussed in the previous subsection. Let us assume that for both switch positions A and B of the PWM-controlled circuit the requirements for the application of Proposition 3.1 are met. Since, in general, for the two switch positions A and B the state-tree representation (3.40) and (3.41) is different, an upper index A or B will subsequently always refer to the corresponding quantity of the PWM-controlled circuit for the s-switch tuple $\{T_1, \ldots, T_s\}$ in position A or B, respectively. Since we assumed in the previous subsection that the expressions for $\tilde{\psi}^j$, $j \in L_i$ and $\tilde{q}_j$, $j \in C_i$ from (3.35) do not depend on the switch position, we may deduce that in (3.40) and (3.41) the relations $\bar{D}_{LL}^A = \bar{D}_{LL}^B = \bar{D}_{LL}$ and $\bar{D}_{CC}^A = \bar{D}_{CC}^B = \bar{D}_{CC}$ hold. This assumption also implies that the expression of the total stored energy in the inductors and capacitors $\hat{w} = \hat{w}_L + \hat{w}_C$ is independent of the switch position.

Based on (3.89), we may apply Proposition 3.1 to the PWM-controlled circuit and the average model can be written as

$$
\frac{d}{dt}x = \left(J^B - S^B + \left((J^A - J^B) - (S^A - S^B)\right)d_A\right)\left(\frac{\partial \hat{w}}{\partial x}\right)^T +
$$
$$
\left(G_{U_0}^B + (G_{U_0}^A - G_{U_0}^B)d_A\right)\begin{bmatrix} 0 \\ U_0^T \end{bmatrix} + \left(G_{I_0}^B + (G_{I_0}^A - G_{I_0}^B)d_A\right)\begin{bmatrix} 0 \\ I_0 \end{bmatrix} \tag{3.90}
$$

with the state (see Remark 3.3) $x^T = \left[\tilde{\psi}_{C,L}, \tilde{q}_{T,C}^T\right]$, $\tilde{\psi}_{C,L} = \psi_{C,L} + \psi_{T,L}\bar{D}_{LL}$, $\tilde{q}_{T,C} = q_{T,C} - \bar{D}_{CC}q_{C,C}$, the duty ratio d_A and J^A, J^B, S^A, S^B, $G_{U_0}^A$, $G_{U_0}^B$, $G_{I_0}^A$ and $G_{I_0}^B$ from (3.44) and (3.45) for the s-switch tuple in position A or B respectively. As one can see from (3.90), depending on the topological location of the s-switch tuple, the duty ratio d_A as the control input has an effect on the internal energy-preserving interconnection part J, the dissipative part S, as well as on the energy flow with the system environment via the independent voltage and current sources, described by G_{U_0} and G_{I_0} respectively. But in general, the duty ratio d_A does not influence all these parts simultaneously, and therefore we will go on to discuss three special cases.

Case I: Influence on the Energy-preserving Part. If the power flow of the static elements remains unchanged from one switch position to the other, then in (3.89) the relations $\hat{p}_S^A = \hat{p}_S^B = \hat{p}_S$ and $\check{p}_S^A = \check{p}_S^B = \check{p}_S$ hold. Following (3.54) and (3.59) of the proof of Proposition 3.1, we immediately see that in this case the duty ratio d_A influences only the internal energy-preserving interconnection part, given by J (compare also with (1.76) of Chapter 1). Consequently, (3.90) simplifies to

$$
\frac{d}{dt}x = \left(J^B + (J^A - J^B)d_A - S\right)\left(\frac{\partial \hat{w}}{\partial x}\right)^T + G_{U_0}\begin{bmatrix} 0 \\ U_0^T \end{bmatrix} + G_{I_0}\begin{bmatrix} 0 \\ I_0 \end{bmatrix}. \tag{3.91}
$$

A representative of this group is the so-called Ćuk-converter [97], an indirect high-frequency switch-mode dc-to-dc converter. Since the next section is devoted to the design of a non-linear H_2-controller with and without integral term for the Ćuk-converter, we will subsequently derive the mathematical model in more detail. Fig. 3.4 depicts the circuit scheme of the Ćuk-converter with the supply voltage U_0, the internal resistances R_1 and R_2 of the inductors L_1 and L_2, the capacitors C_1, C_2 and the load conductance G_L. Variations of the load will be considered in the sense of a Norton equivalent circuit [57] in the form of changes in the output current Δi_o. Fluctuations of the input voltage will be denoted by ΔU_0. The Ćuk-converter has one switch T_1 with the on-position A and the off-position B. It is assumed that the switch T_1 has ideal characteristics, that means no losses and zero turn-on and turn-off times. One can easily see that for the Ćuk-converter in fact $\hat{p}_S$ and $\check{p}_S$ are

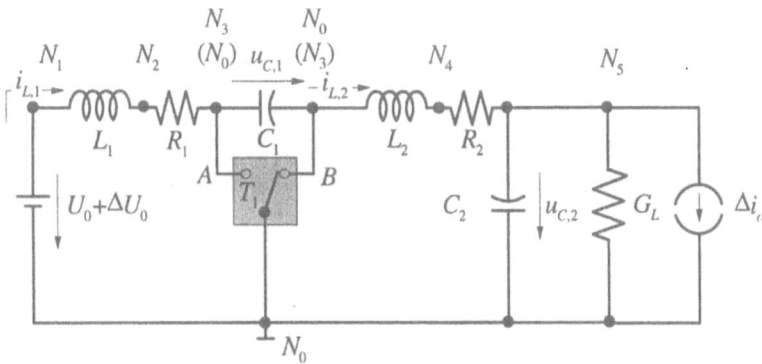

Fig. 3.4. Circuit scheme of the Ćuk-converter.

independent of the switch position A and B, namely

$$\check{p}_S^A = \check{p}_S^B = \check{p}_S = \frac{1}{2}\left(R_1 i_{L,1}^2 + R_2 i_{L,2}^2 + u_{C,2}^2 G_L\right) - U_0 i_{L,1}$$
$$\hat{p}_S^A = \hat{p}_S^B = \hat{p}_S = \frac{1}{2}\left(R_1 i_{L,1}^2 + R_2 i_{L,2}^2 + u_{C,2}^2 G_L\right) .$$

(3.92)

Fig. 3.5 shows the digraph associated with the Ćuk-converter for the switch T_1 in position B and a tree indicated by the dotted lines for the state-tree representation. The orientation of the branches corresponds with the associated reference direction of the current flows. Using the state-tree of Fig. 3.5, we can write (3.40), for the Ćuk-converter with the switch T_1 in position B, in the form

$$\begin{bmatrix} i_{C,1} \\ i_{C,2} \\ i_{R,1} \\ i_{R,2} \\ i_{U,0} \end{bmatrix} = \begin{bmatrix} 1 & 0 & 0 & 0 \\ 0 & 1 & -1 & -1 \\ 1 & 0 & 0 & 0 \\ 0 & 1 & 0 & 0 \\ 1 & 0 & 0 & 0 \end{bmatrix} \begin{bmatrix} i_{L,1} \\ i_{L,2} \\ i_{G,L} \\ \Delta i_o \end{bmatrix} .$$

(3.93)

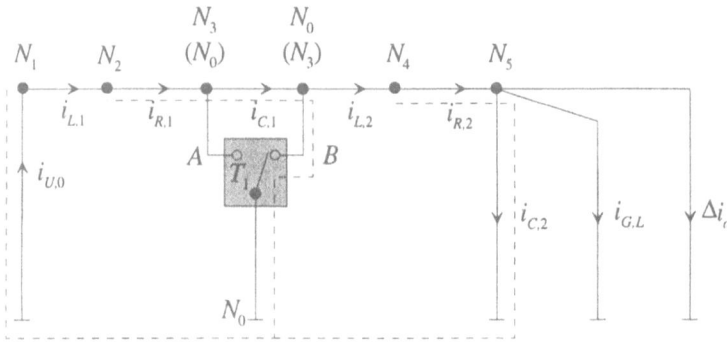

Fig. 3.5. Digraph and state-tree for Fig. 3.4 with switch T_1 in position B.

By means of a suitable partitioning of (3.93) Proposition 3.1 leads to the mathematical model of the Ćuk-converter for the switch T_1 in position B

$$\frac{d}{dt}x = \left(\begin{bmatrix} 0 & \Lambda^B \\ -\left(\Lambda^B\right)^T & 0 \end{bmatrix} - \begin{bmatrix} S_1 & 0 \\ 0 & S_2 \end{bmatrix} \right) \left(\frac{\partial \hat{w}}{\partial x}\right)^T + G_e \begin{bmatrix} U_0 + \Delta U_0 \\ \Delta i_o \end{bmatrix}$$

(3.94)

with the state $x^T = [L_1 i_{L,1},\ L_2 i_{L,2},\ C_1 u_{C,1},\ C_2 u_{C,2}]$, the total energy stored in the inductors and capacitors

$$\hat{w} = \frac{1}{2}\left(L_1 i_{L,1}^2 + L_2 i_{L,2}^2 + C_1 u_{C,1}^2 + C_2 u_{C,2}^2 \right)$$

(3.95)

and the matrices

$$\Lambda^B = \begin{bmatrix} -1 & 0 \\ 0 & -1 \end{bmatrix},\ S_1 = \begin{bmatrix} R_1 & 0 \\ 0 & R_2 \end{bmatrix},\ S_2 = \begin{bmatrix} 0 & 0 \\ 0 & G_L \end{bmatrix},\ G_e = \begin{bmatrix} 1 & 0 \\ 0 & 0 \\ 0 & 0 \\ 0 & -1 \end{bmatrix}.$$

(3.96)

As discussed before the mathematical model of the Ćuk-converter for the switch T_1 in position A differs from (3.94) only in the energy-preserving part, i.e. Λ^B has to be replaced by Λ^A,

$$\Lambda^A = \begin{bmatrix} 0 & 0 \\ -1 & -1 \end{bmatrix}.$$

(3.97)

Thus, following (3.91), we can write the average model of the Ćuk-converter in the form

$$\frac{d}{dt}x = \left(J^B + \left(J^A - J^B \right) d_A - S \right) \left(\frac{\partial \hat{w}}{\partial x}\right)^T + G_e \begin{bmatrix} U_0 + \Delta U_0 \\ \Delta i_o \end{bmatrix}$$

(3.98)

with the state (see Remark 3.3) $x^T = [L_1 i_{L,1}, L_2 i_{L,2}, C_1 u_{C,1}, C_2 u_{C,2}]$, the duty ratio d_A, the total energy stored in the system due to (3.95) and

$$
J^B = \begin{bmatrix} 0 & 0 & -1 & 0 \\ 0 & 0 & 0 & -1 \\ 1 & 0 & 0 & 0 \\ 0 & 1 & 0 & 0 \end{bmatrix}, \quad J^A = \begin{bmatrix} 0 & 0 & 0 & 0 \\ 0 & 0 & -1 & -1 \\ 0 & 1 & 0 & 0 \\ 0 & 1 & 0 & 0 \end{bmatrix}, \quad S = \begin{bmatrix} R_1 & 0 & 0 & 0 \\ 0 & R_2 & 0 & 0 \\ 0 & 0 & 0 & 0 \\ 0 & 0 & 0 & G_L \end{bmatrix}, \quad (3.99)
$$

$$
G_e^T = \begin{bmatrix} 1 & 0 & 0 & 0 \\ 0 & 0 & 0 & -1 \end{bmatrix}. \tag{3.100}
$$

Case II: Dissipative Breaking. Remember that from (3.49) we can subdivide $\hat{p}_S$ and $\check{p}_S$ into parts due to the independent voltage and current sources, $\hat{p}_S^{U_0}$ and $\check{p}_S^{I_0}$, and into parts due to the resistors, $\hat{p}_S^R$ and $\check{p}_S^R$, i.e. $\hat{p}_S = \hat{p}_S^R + \hat{p}_S^{U_0}$ and $\check{p}_S = \check{p}_S^R + \check{p}_S^{I_0}$. Let us consider PWM-controlled electric circuits where in (3.89) only the expressions of $\hat{p}_S^R$ and $\check{p}_S^R$ change with the switch position. Then, following (3.68) and (3.71) of the proof of Proposition 3.1, we can immediately see that (3.90) simplifies to

$$
\begin{aligned}
\frac{\mathrm{d}}{\mathrm{dt}} x = {} & \left(J - S^B - (S^A - S^B) d_A \right) \left(\frac{\partial \hat{w}}{\partial x} \right)^T \\
& + \begin{bmatrix} G_{U_0,L} \\ \left(G_{U_0,C}^B + (G_{U_0,C}^A - G_{U_0,C}^B) d_A \right) \end{bmatrix} \begin{bmatrix} 0 \\ U_0^T \end{bmatrix} \\
& + \begin{bmatrix} \left(G_{I_0,L}^B + (G_{I_0,L}^A - G_{I_0,L}^B) d_A \right) \\ G_{I_0,C} \end{bmatrix} \begin{bmatrix} 0 \\ I_0 \end{bmatrix}
\end{aligned} \tag{3.101}
$$

with G_{U_0} and G_{I_0} from (3.45). In this case, when the duty ratio d_A influences the dissipative part, the system is also called PWM-controlled with dissipative breaking. Fig. 3.6 shows a very simple representative of this group, with $R_{L,1} \neq R_{L,2}$.

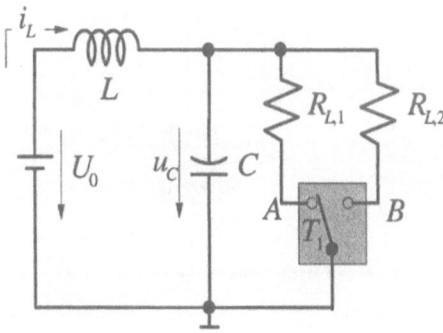

Fig. 3.6. Simple PWM-controlled circuit with dissipative breaking.

The associated mathematical model reads as

$$\frac{d}{dt}x = \left(J - S^B - \left(S^A - S^B\right)d_A\right)\left(\frac{\partial \hat{w}}{\partial x}\right)^T + G_e U_0 \tag{3.102}$$

with the state $x^T = [Li_L, Cu_C]$, the duty ratio d_A, the energy stored in the system $\hat{w} = \frac{1}{2}Li_L^2 + \frac{1}{2}Cu_C^2$ and

$$J = \begin{bmatrix} 0 & -1 \\ 1 & 0 \end{bmatrix}, \ S^B = \begin{bmatrix} 0 & 0 \\ 0 & \frac{1}{R_{L,2}} \end{bmatrix}, \ S^A = \begin{bmatrix} 0 & 0 \\ 0 & \frac{1}{R_{L,1}} \end{bmatrix}, \ G_e = \begin{bmatrix} 1 \\ 0 \end{bmatrix}. \tag{3.103}$$

Case III: Influence on the Energy Flow with the System Environ-ment. Here, we consider PWM-controlled electric circuits where in (3.89) only the expressions of $\hat{p}_S^{U_0}$ and $\check{p}_S^{I_0}$ change with the switch position. Follow-ing (3.64) and (3.65) of the proof of Proposition 3.1, we can write (3.90) in the form

$$\frac{d}{dt}x = (J - S)\left(\frac{\partial \hat{w}}{\partial x}\right)^T + \left[\begin{array}{c} \left(G_{U_0,L}^B + \left(G_{U_0,L}^A - G_{U_0,L}^B\right)d_A\right) \\ G_{U_0,C} \end{array}\right]\begin{bmatrix} 0 \\ U_0^T \end{bmatrix}$$

$$+ \left[\begin{array}{c} G_{I_0,L} \\ \left(G_{I_0,C}^B + \left(G_{I_0,C}^A - G_{I_0,C}^B\right)d_A\right) \end{array}\right]\begin{bmatrix} 0 \\ I_0 \end{bmatrix}$$

$$\tag{3.104}$$

with G_{U_0} and G_{I_0} from (3.45). As an example, in Fig. 3.7 a simple dc-to-dc converter with a 4-switch tuple $\{T_1, T_2, T_3, T_4\}$ is presented.

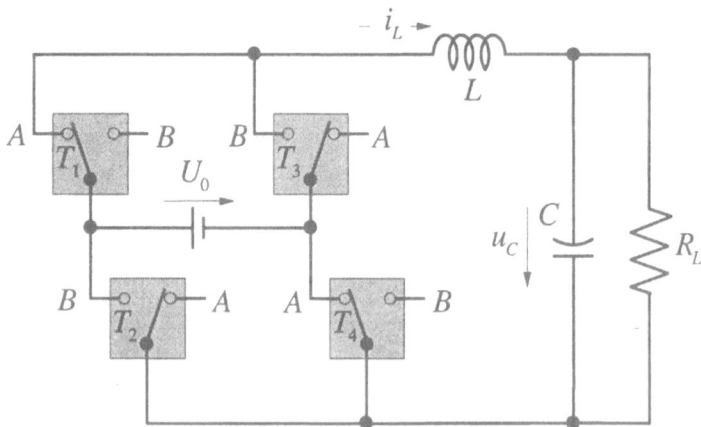

Fig. 3.7. Simple dc-to-dc converter with four switches.

The mathematical model takes the form

$$\frac{\mathrm{d}}{\mathrm{d}t}x = (J - S)\left(\frac{\partial \hat{w}}{\partial x}\right)^T + \left[\begin{array}{c} \left(G^B_{U_0,L} + (G^A_{U_0,L} - G^B_{U_0,L})\, d_A\right) \\ G_{U_0,C} \end{array} \right] U_0 \quad (3.105)$$

with the state $x^T = [Li_L, Cu_C]$, the duty ratio d_A, the energy stored in the system $\hat{w} = \frac{1}{2}Li_L^2 + \frac{1}{2}Cu_C^2$ and

$$J = \begin{bmatrix} 0 & -1 \\ 1 & 0 \end{bmatrix}, \; S = \begin{bmatrix} 0 & 0 \\ 0 & \dfrac{1}{R_L} \end{bmatrix}, \; G^B_{U_0,L} = -1, \; G^A_{U_0,L} = 1 \;. \quad (3.106)$$

3.5 Application: Non-linear Control of a Ćuk-converter

The Ćuk-converter is an indirect high-frequency switch-mode dc-to-dc converter. Generally, dc-to-dc converters are used as interfaces between dc systems of different voltage levels in regulated power supplies for different types of electronic equipment and in dc-motor drive applications (see, e.g., [57], [99]). By means of feedback control the average dc-output voltage of the dc-to-dc converter must be controlled to a desired level and the dc-output must be kept at this level if there are any variations of the load or fluctuations in the input voltage. In so-called high-frequency switch-mode converters the average output voltage is controlled by adjusting the on- and off-durations of a semiconductor device, switching at a rate that is fast compared to the changes of the input and output signals. In the PWM-case this switching frequency is constant and here the ratio of the on-duration of the switch to the fixed switching time period, also denoted as the duty ratio, is used for controlling the system. Within the high-frequency switch-mode dc-to-dc converters, there is an additional classification, namely the direct and the indirect converters [57]. Direct converters have a direct dc path between the input and output port. They are known as either a down (buck) converter or an up (boost) converter, depending on the direction of power flow. The indirect converter topologies have no direct dc path between the input and output ports in any switch state. The best-known examples are the up/down (buck/boost) converter and the Ćuk-converter [97]. These converter types have the following properties: the input and output voltages are of opposite signs and the output can be either higher or lower than the input for both the voltage and the current. The Ćuk-converter uses two inductors at the input and output port with the advantage that the switching frequency ripple in the input and output current is reduced to an acceptable level.

The modeling and control of dc-to-dc power converters have been studied for many years and the results are reported in various textbooks and journals. Beside the traditionally used linear approaches (e.g., [57], [79]), which suffer from the lack of a stability proof, different non-linear design methods are presented in [128] - [131] and the references cited there. In [128] and [130] a non-linear PI (proportional integral) controller is designed on the basis of extended linearization techniques, using a family of linearized plant models parametrized by equilibrium points. Differential geometric methods for output regulation, like the input-output linearization (see, e.g., [52], [103], [144]), cannot be applied in a straightforward way to indirect converter types like the Ćuk-converter due to the unstable zero-dynamics. In order to avoid stability problems, output regulation can be achieved only indirectly by regulation of other state variables, for example the input current (see, e.g., [129], [131]). Since the output voltage is controlled only indirectly by such a strategy however, it requires additional effort to take variations of the load into consideration. Besides the classical PWM techniques the theory of VSS (Variable Structure Systems) with its associated sliding regimes has been proposed as a means of designing the controller for dc-to-dc power converters (see, e.g., [113]). The reader is also encouraged to read a very important recent book [107], where a Lagrangian and Hamiltonian dynamics approach (see, also [26], [132]) is used to model switched dc-to-dc power converters. Here, also, the physical properties of these models are used to advantage to design passivity-based feedback controllers. The importance of this passivity-based concept is the fact that the controller design philosophy is not only mathematically motivated, it also takes advantage of the physical structure of the system to be controlled. Subsequently, the presented controller design for the Ćuk-converter is based on the non-linear state feedback H_2-design with and without integral term of Section 2.1 and 2.2 (see, also [66], [72] for a non-linear H_∞-approach).

3.5.1 Mathematical Model

Consider the Ćuk-converter of Fig. 3.4 with the average model (3.98) - (3.100). Let us at first assume that there are no load variations and no fluctuations in the input voltage, i.e. $\Delta i_o = 0$ and $\Delta U_0 = 0$. Further, suppose that $\bar{d}_A$ determines the operating point $\bar{x}^T = [L_1 \bar{i}_{L,1}, L_2 \bar{i}_{L,2}, C_1 \bar{u}_{C,1}, C_2 \bar{u}_{C,2}]$ of the Ćuk-converter with

$$
\begin{aligned}
\bar{i}_{L,1} &= \frac{G_L U_0 \bar{d}_A^2}{(1 + R_2 G_L)\left(1 - \bar{d}_A\right)^2 + \bar{d}_A^2 G_L R_1} \\
\bar{i}_{L,2} &= \frac{-G_L U_0 \bar{d}_A \left(1 - \bar{d}_A\right)}{(1 + R_2 G_L)\left(1 - \bar{d}_A\right)^2 + \bar{d}_A^2 G_L R_1} \\
\bar{u}_{C,1} &= \frac{U_0 \left(1 + G_L R_2\right)\left(1 - \bar{d}_A\right)}{(1 + R_2 G_L)\left(1 - \bar{d}_A\right)^2 + \bar{d}_A^2 G_L R_1} \\
\bar{u}_{C,2} &= \frac{-U_0 \bar{d}_A \left(1 - \bar{d}_A\right)}{(1 + R_2 G_L)\left(1 - \bar{d}_A\right)^2 + \bar{d}_A^2 G_L R_1} .
\end{aligned} \tag{3.107}
$$

Then, by means of a simple change of coordinates $x = \Delta x + \bar{x}$ this operating point is shifted to the origin and with $d_A = \Delta d_A + \bar{d}_A$ the system (3.98) can be rewritten in the form

$$\frac{d}{dt}\Delta x = (\bar{J} - S)\left(\frac{\partial \hat{w}_\Delta}{\partial \Delta x}\right)^T + (J^A - J^B)\left(\frac{\partial \hat{w}_\Delta}{\partial \Delta x} + \frac{\partial \hat{w}}{\partial x}(\bar{x})\right)^T \Delta d_A,$$

(3.108)

where $\bar{J} = J^B + \left(J^A - J^B\right)\bar{d}_A$ and $\hat{w}_\Delta$ is the shifted energy function

$$\hat{w}_\Delta = \frac{1}{2}\left(L_1 \Delta i_{L,1}^2 + L_2 \Delta i_{L,2}^2 + C_1 \Delta u_{C,1}^2 + C_2 \Delta u_{C,2}^2\right).$$

(3.109)

At this point it is worth mentioning that (3.108) can also be obtained by means of the mixed potentials of the shifted system $p_{C,\Delta}^K$, $\hat{p}_{S,\Delta}^K$, $p_{L,\Delta}^K$ and $\check{p}_{S,\Delta}^K$, $K = A, B$ (see Theorem 3.2 and 3.3). But now $\hat{p}_{S,\Delta}^K$ and $\check{p}_{S,\Delta}^K$ are no longer independent of Δd_A and the change of the shifted energy function $\hat{w}_\Delta$ meets the relation

$$\frac{d}{dt}\hat{w}_\Delta = -\left(\hat{p}_{S,\Delta}^B + \check{p}_{S,\Delta}^B\right) - \left(\hat{p}_{S,\Delta}^A - \hat{p}_{S,\Delta}^B + \check{p}_{S,\Delta}^A - \check{p}_{S,\Delta}^B\right)\Delta d_A$$

$$= -\left(\frac{\partial \hat{w}_\Delta}{\partial \Delta x}\right)S\left(\frac{\partial \hat{w}_\Delta}{\partial \Delta x}\right)^T + \left(\frac{\partial \hat{w}_\Delta}{\partial \Delta x}\right)(J^A - J^B)\left(\frac{\partial \hat{w}}{\partial x}(\bar{x})\right)^T \Delta d_A.$$

(3.110)

From (3.110) and LaSalle's invariance principle we can immediately see that the equilibrium $\Delta x = 0$ of the free system (3.108), i.e. $\Delta d_A = 0$, is asymptotically stable in the sense of Lyapunov and the shifted energy function $\hat{w}_\Delta$ from (3.109) serves as an appropriate Lyapunov function.

3.5.2 Non-linear State Feedback H_2-design

Let us consider the mathematical model of the Ćuk-converter (3.108) with the plant output $y = \Delta u_{C,2}$. Following Section 2.1, we are looking for a control law $\Delta d_A = \Delta d_A(\Delta x)$, $\Delta d_A(0) = 0$ such that the origin is rendered asymptotically stable and the objective function

$$J_2 = \inf_{\Delta d_A \in L_2[0,\infty)} \frac{1}{2}\int_0^\infty \left(G_L \Delta u_{C,2}^2 + \Delta d_A^2\right) dt$$

(3.111)

is minimized with respect to Δd_A. Note that we have chosen the nominal load conductance G_L as the weighting factor β for the output function (compare with (2.4)). Following Theorem 2.1, we have to find a C^1 positive definite solution $V(\Delta x)$ of the associated HJBi (2.11). As a candidate for $V(\Delta x)$ we take the shifted energy function $\hat{w}_\Delta$ of (3.109) and the HJBi becomes

$$-\left(\frac{\partial \hat{w}_\Delta}{\partial \Delta x}\right)S\left(\frac{\partial \hat{w}_\Delta}{\partial \Delta x}\right)^T + \frac{1}{2}G_L \Delta u_{C,2}^2 - \frac{1}{2}\left(\Delta d_A^*\right)^2 \leq 0$$

(3.112)

with Δd_A^* as the optimal control law due to (2.9)

$$\Delta d_A^* = -\left(\frac{\partial \hat{w}_\Delta}{\partial \Delta x}\right)\left(J^A - J^B\right)\left(\frac{\partial \hat{w}}{\partial x}(\bar{x})\right)^T \tag{3.113}$$
$$= -\left(\Delta u_{C,1}\left(\bar{i}_{L,2} - \bar{i}_{L,1}\right) - \bar{u}_{C,1}\left(\Delta i_{L,2} - \Delta i_{L,1}\right)\right).$$

By substituting S from (3.99) into (3.112), we get

$$-R_1 \Delta i_{L,1}^2 - R_2 \Delta i_{L,2}^2 - \frac{1}{2} G_L \Delta u_{C,2}^2 - \frac{1}{2}\left(\Delta d_A^*\right)^2 \le 0 \tag{3.114}$$

and hence the HJBi is, in fact, satisfied. Furthermore, it is easy to check that the system

$$\frac{\mathrm{d}}{\mathrm{d}t}\Delta x = \left(J^B + \left(J^A - J^B\right)\bar{d}_A - S\right)\left(\frac{\partial \hat{w}_\Delta}{\partial \Delta x}\right)^T$$
$$\Delta \bar{y}^T = \left[\Delta u_{C,2}, \left(\frac{\partial \hat{w}_\Delta}{\partial \Delta x}\right)\left(J^A - J^B\right)\left(\frac{\partial \hat{w}}{\partial x}(\bar{x})\right)^T\right] \tag{3.115}$$

is zero-state observable. Thus, from Theorem 2.1, we may conclude that the control law (3.113) solves the suboptimal non-linear H_2-design problem with respect to the objective function (3.111) for the Ćuk-converter model (3.108) and the output $y = \Delta u_{C,2}$.

3.5.3 Non-linear State Feedback H_2-design with Integral Term

The non-linear H_2-controller (3.113) does not show a satisfactory performance in the case when there are changes of the operating point due to load variations or fluctuations in the supply voltage. To overcome these deficiencies we want to add an integral term to the H_2-controller. This can be done by applying Proposition 2.1 to the Ćuk-converter model. Therefore, in a first step, we rewrite the mathematical model of the Ćuk-converter (3.108) and (3.109) according to (2.26) in the form

$$\frac{\mathrm{d}}{\mathrm{d}t}x = Ax + b\left(x\right)u$$
$$y = c^T x \tag{3.116}$$

with the state $x^T = [\Delta i_{L,1}, \Delta i_{L,2}, \Delta u_{C,1}, \Delta u_{C,2}]$, the plant input $u = \Delta d_A$, the plant output y, the vectors $c^T = [0, 0, 0, 1]$,

$$b^T\left(x\right) = \left[\frac{\Delta u_{C,1} + \bar{u}_{C,1}}{L_1}, \frac{-\Delta u_{C,1} - \bar{u}_{C,1}}{L_2}, \frac{\Delta i_{L,2} + \bar{i}_{L,2} - \Delta i_{L,1} - \bar{i}_{L,1}}{C_1}, 0\right] \tag{3.117}$$

and the matrix

$$
A = \begin{bmatrix}
-\dfrac{R_1}{L_1} & 0 & \dfrac{\bar{d}_A - 1}{L_1} & 0 \\[2mm]
0 & \dfrac{-R_2}{L_2} & \dfrac{-\bar{d}_A}{L_2} & \dfrac{-1}{L_2} \\[2mm]
\dfrac{1 - \bar{d}_A}{C_1} & \dfrac{\bar{d}_A}{C_1} & 0 & 0 \\[2mm]
0 & \dfrac{1}{C_2} & 0 & \dfrac{-G_L}{C_2}
\end{bmatrix}.
\tag{3.118}
$$

It can be easily seen that for $0 < \bar{d}_A < 1$ the matrix A is Hurwitz and the pair (A, c) is observable. Furthermore, the condition $c^T A^{-1} b\,(0) \neq 0$ is satisfied for all $\bar{d}_A$ except for the duty ratio

$$
\bar{d}_A = \frac{1}{2} \frac{-2R_2 G_L - 2 + 2\sqrt{R_2 G_L^2 R_1 + G_L R_1}}{-R_2 G_L + G_L R_1 - 1}.
\tag{3.119}
$$

Notice that the stationary duty ratio of (3.119) is exact that value where the stationary output voltage $\bar{u}_{C,2}$ of (3.107) reaches its maximum for given parameters R_1, R_2, G_L and U_0. Due to reasons of efficiency all operating points with a stationary duty ratio $\bar{d}_A$ greater or equal than the value of (3.119) are not feasible (see also the discussion at the beginning of the next subsection, in particular Fig. 3.9). Hence, all the requirements for the application of Proposition 2.1 are met. Thus, the non-linear state feedback controller with integral term can be obtained directly from (2.29) and (2.30) of Proposition 2.1.

Since the symbolic expressions of this control law are too big to be presented here, we will set the internal resistances R_1 and R_2 of the inductors L_1 and L_2 to zero. This simplification does not change anything for the controller design, it just makes the expressions of the non-linear controller more readable. Then the unique positive definite solution P_{11} of the Lyapunov equation (2.30) is given by the shifted energy function (3.109)

$$
P_{11} = \frac{\beta}{G_L} \hat{w}_\Delta = \frac{1}{2} \frac{\beta}{G_L} \left(L_1 \Delta i_{L,1}^2 + L_2 \Delta i_{L,2}^2 + C_1 \Delta u_{C,1}^2 + C_2 \Delta u_{C,2}^2 \right).
\tag{3.120}
$$

Consequently, the non-linear H_2-controller with integral term due to (2.29) reads as

$$
\frac{\mathrm{d}}{\mathrm{dt}} x_I = \Delta u_{C,2}
$$

$$
u = -\frac{\beta}{2 G_L} \left(\Delta u_{C,1} \left(\bar{i}_{L,2} - \bar{i}_{L,1} \right) - \bar{u}_{C,1} \left(\Delta i_{L,2} - \Delta i_{L,1} \right) \right) -
$$

$$
p_{22} \frac{\left(\bar{u}_{C,1} + \Delta u_{C,1} \right)}{\left(1 - \bar{d}_A \right)^2} \left(\bar{d}_A L_1 \Delta i_{L,1} - \left(1 - \bar{d}_A \right) L_2 \Delta i_{L,2} - \left(1 - \bar{d}_A \right) x_I \right)
\tag{3.121}
$$

with the controller parameters β, $p_{22} > 0$. Notice that the part of the controller (3.121) weighted with β is identical with the pure non-linear H_2-controller (3.113). The controller parameters β and p_{22} are used to adjust the performance of the closed-loop, where the choice of these parameters results from the following heuristic considerations. The parameter p_{22} has a direct influence on the integral action and by means of an increasing β more damping is injected into the system.

3.5.4 The Experimental Setup

Fig. 3.8 shows the laboratory model for performing the Ćuk-converter experiments with the parameter values $L_1 = L_2 = 10.9 \cdot 10^{-3}$ H, $R_1 = R_2 = 1.3$ Ω, $C_1 = 22.0 \cdot 10^{-6}$ F, $C_2 = 22.9 \cdot 10^{-6}$ F and $U_0 = 12$ V ([2], [45]). The capacitor C_1 is located in an external pin base and can also be exchanged as required by the experiments. The load can be chosen to be either a resistor with a fixed conductance $G_L = 1/22.36$ S, or the load conductance can be

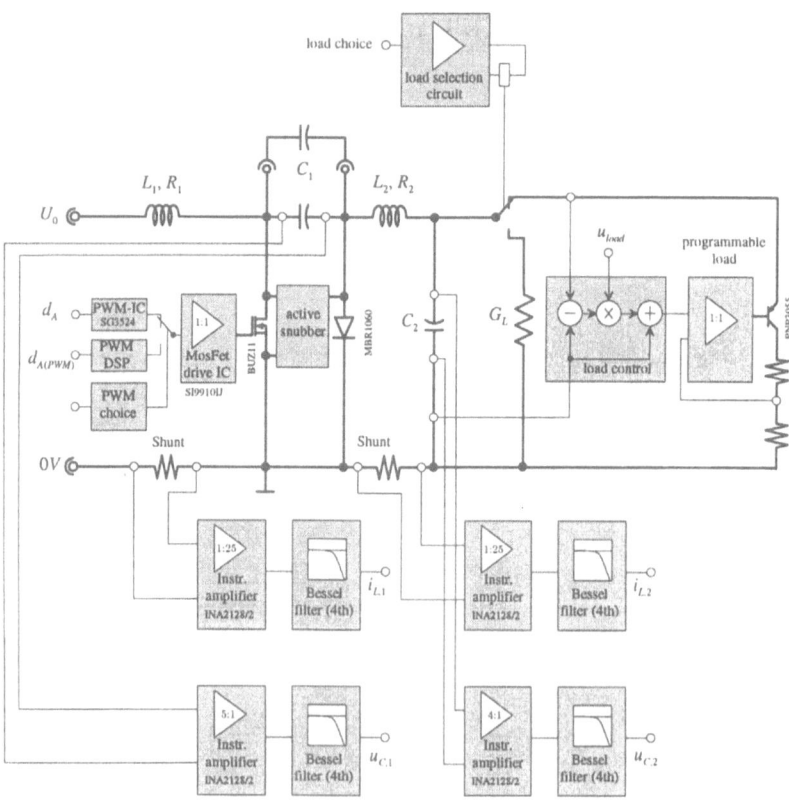

Fig. 3.8. Block diagram for the experimental setup.

set to an arbitrary value in a range $G_L \leq 1/6.4$ S via a programmable load simulator. The value of the programmable load simulator can be defined by a control voltage u_{load}.

Fig. 3.9, left hand side, shows the achievable stationary output voltage $\bar{u}_{C,2}$ as a function of the stationary duty ratio $\bar{d}_A$ and the load conductance G_L. As one can see for a fixed load resistor G_L^{-1}, two different values of the duty ratio $\bar{d}_A$ exist to obtain a certain output voltage $\bar{u}_{C,2}$. Due to reasons of efficiency the duty ratio $\bar{d}_A$ with the lower value is only feasible. This is why the stationary duty ratio is restricted to an upper bound which depends on the load conductance G_L. Fig. 3.9, right hand side, presents the maximum possible stationary duty ratio $\bar{d}_{A,\max}$ as a function of the load resistor G_L^{-1}. It is worth mentioning that the effect presented in Fig. 3.9 vanishes if the

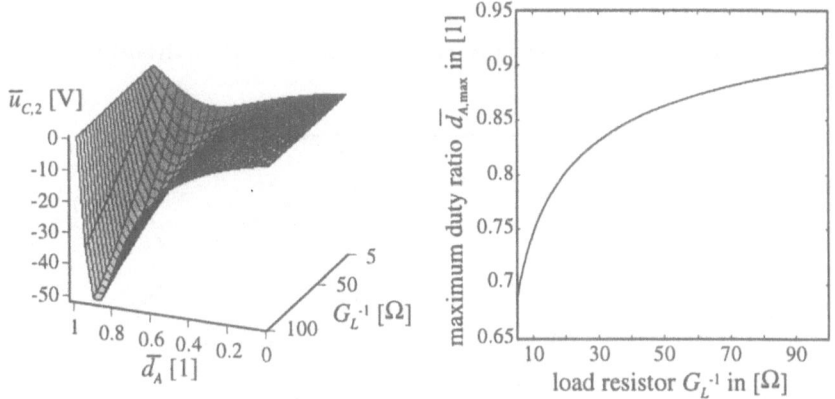

Fig. 3.9. Left hand side: Stationary output voltage $\bar{u}_{C,2}$ as a function of the duty ratio $\bar{d}_A$ and the load resistor G_L^{-1}. Right hand side: Maximum possible stationary duty ratio $\bar{d}_{A,\max}$ as a function of the load resistor G_L^{-1}.

internal resistances R_1 and R_2 are zero. This can be easily seen because then the stationary output voltage $\bar{u}_{C,2}$ as a function of $\bar{d}_A$ reads as

$$\bar{u}_{C,2} = \frac{-\bar{d}_A U_0}{\left(1 - \bar{d}_A\right)} \ . \qquad (3.122)$$

The switch S is realized with a standard MOSFET (BUZ11) in combination with the MOSFET-drive-IC SI9910DJ and a Shottky Diode (MBR1060) with a low forward voltage drop. An active turn-off snubber is used to provide a zero voltage across the MOSFET, while the current turns off and thus guarantees that the MOSFET is operating within the SOA (safe operating area). In contrast to R-C-D-snubbers, this implemented active snubber does not create additional losses in the system [25]. The modulation frequency for the

PWM actuator (either IC SG3524 or internal PWM of the DSP-unit) is chosen as 25 kHz in order to keep the total losses in the converter to a minimum. The two inductor currents $i_{L,1}$ and $i_{L,2}$ are measured by means of 0.1 Ω shunt resistors and instrumentation amplifier ICs (Burr Brown INA2128/2) with low offset and drift. The capacitor voltages $u_{C,1}$ and $u_{C,2}$ are also directly measured by means of the instrumentation amplifier ICs. All the measurement signals are filtered with fourth order analog Bessel low-pass filters with a cutoff frequency of 10 kHz. The Ćuk-converter experiment operates together with a DSP (digital signal processor)-system (dSpace) integrated in a PC with the operating system WINDOWS NT which enables us to use the MATLAB/SIMULINK environment directly to test the controllers. This hard- and software configuration allows sampling times down to $1 \cdot 10^{-4}$ s. Fig. 3.10 depicts the stationary behavior of the state variables as a function of the duty ratio. The difference between the calculated and the measured curve, especially in the case of $u_{C,1}$, is due to the forward voltage drop of the diode, which is not taken into consideration in the model. For analyzing small-signal

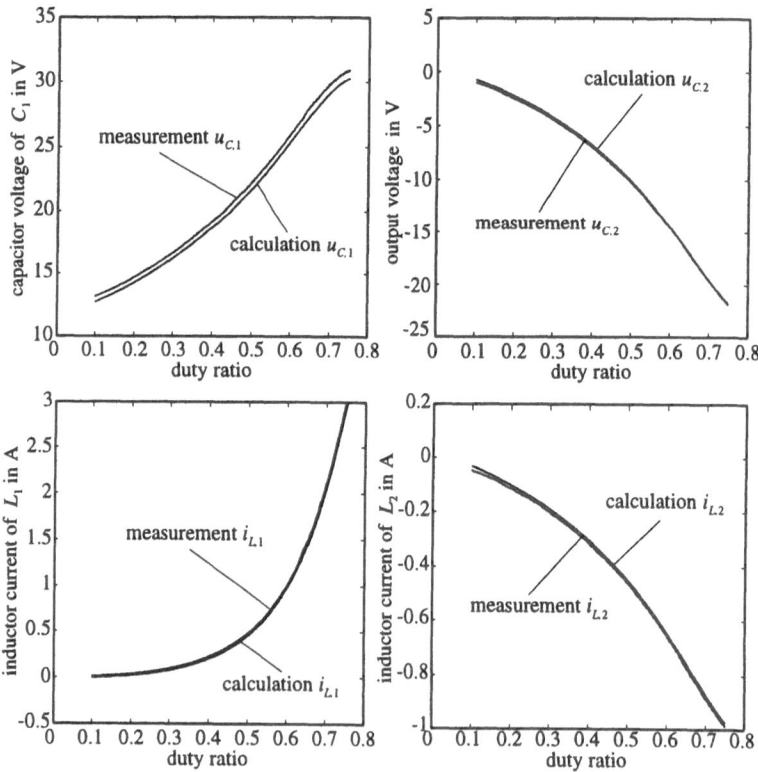

Fig. 3.10. Stationary behavior of the Ćuk-converter.

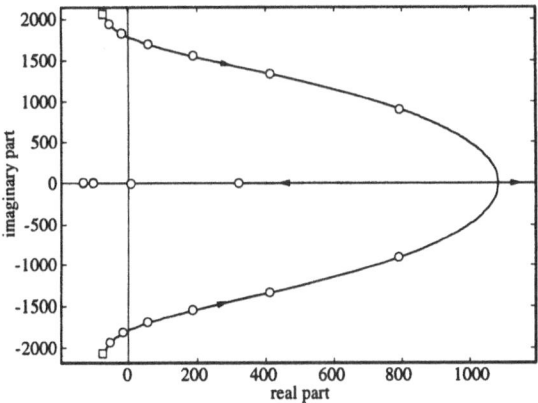

Fig. 3.11. Zeros of the transfer function $Z\left(s, \bar{d}_A\right)$ as a function of the stationary duty ratio $\bar{d}_A$.

dynamics, the Ćuk-converter system (3.98) with $\Delta U_0 = \Delta i_o = 0$ is linearized around an operating point $[\bar{\imath}_{L,1}, \bar{\imath}_{L,2}, \bar{u}_{C,1}, \bar{u}_{C,2}]$ (see (3.107)) and we obtain the mathematical model in terms of small deviations δ around the equilibrium point. Taking δd_A as the plant input and $\delta u_{C,2}$ as the plant output, we can compute the transfer function $Z\left(s, \bar{d}_A\right)$ with the Laplace variable s and the duty ratio $\bar{d}_A$ as a parameter. Now, it is possible to calculate the poles and zeros of the transfer function as a function of $\bar{d}_A$. For the laboratory model the zeros are depicted in Fig. 3.11 and the poles in Fig. 3.12, respectively, whereby the square indicates the point $\bar{d}_A = 0$ and the circles represent the results for $\bar{d}_A$ in 0.1 steps from $\bar{d}_A = 0$ to $\bar{d}_A = 1$. An important feature

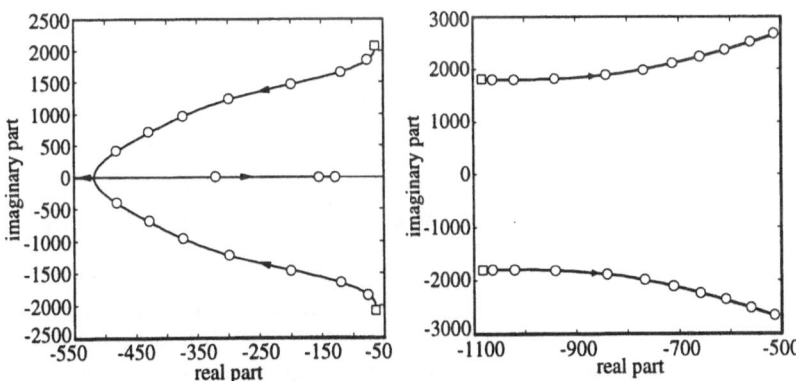

Fig. 3.12. Poles of the transfer function $Z\left(s, \bar{d}_A\right)$ as a function of the stationary duty ratio $\bar{d}_A$.

of the Ćuk-converter is the fact that from $\bar{d}_A = 0.227$ upwards the zeros of the transfer function lie in the closed right half s-plane. In other words the Ćuk-converter shows a bifurcation of the zero dynamics. This fact can also be seen in Fig. 3.13, where the measured and simulated transient responses of the non-linear model for a step input of the duty ratio $\delta d_A = 0.2\sigma\left(t - 5 \cdot 10^{-3}\right)$, for two operating points $\bar{d}_A = 0$ and $\bar{d}_A = 0.5$ are illustrated. In the case of $\bar{d}_A = 0.5$ the step response of $i_{L,2}$ and $u_{C,2}$ show the typical non-minimum phase behavior. Fig. 3.14 demonstrates the transient responses of the Ćuk-

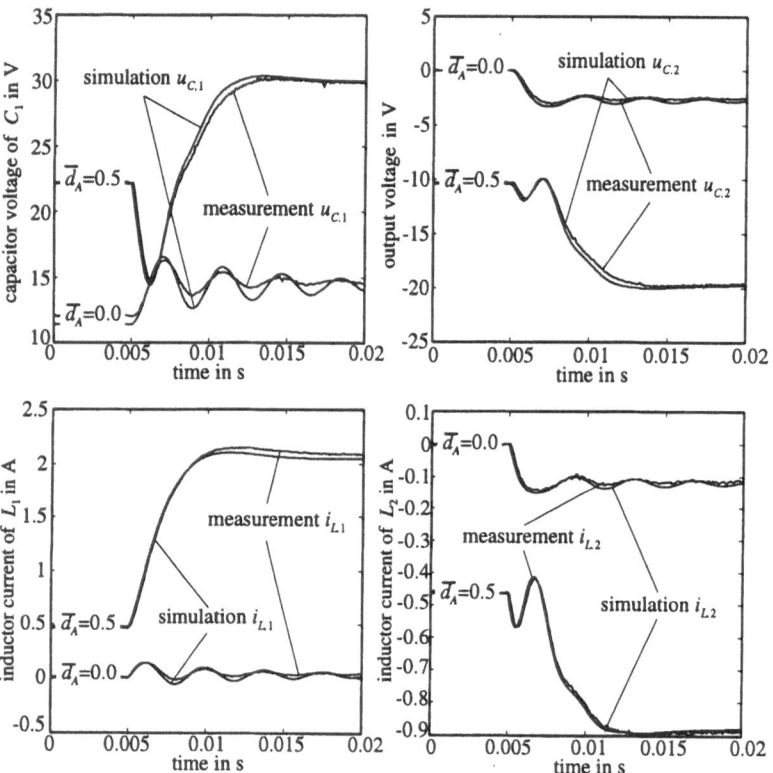

Fig. 3.13. Step responses at two different operating points $\bar{d}_A = 0$ and $\bar{d}_A = 0.5$.

converter in the case of large deviations from the nominal point for a step input of the duty ratio $d_A = 0.5\sigma\left(t - 5 \cdot 10^{-3}\right)$. The difference in the damping behavior between simulation and measurement can be explained by the fact that in the model, the switch (transistor in combination with the diode) is assumed to be ideal, i.e. there are no losses and hence the measured step responses are more damped than the simulated ones.

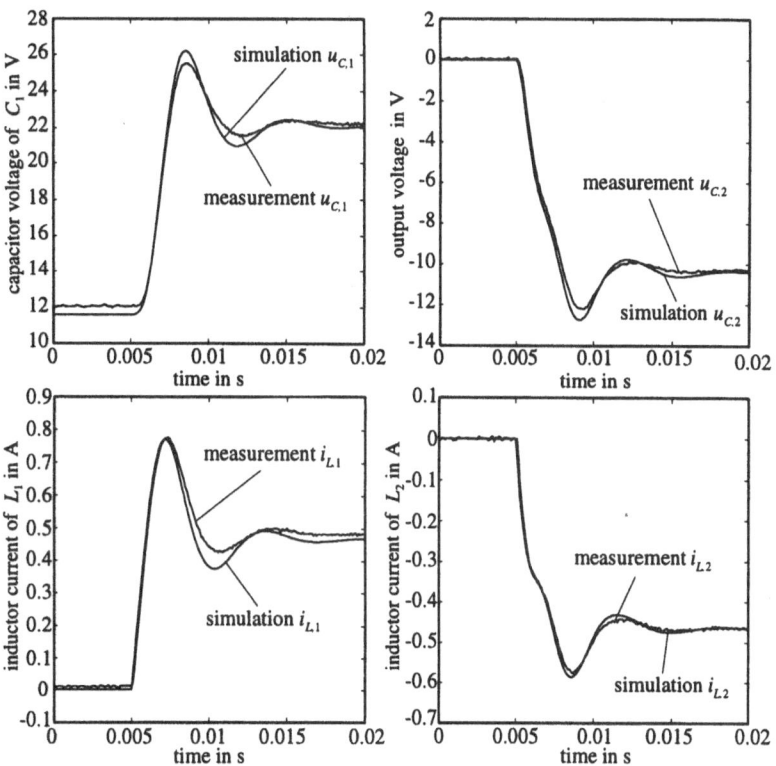

Fig. 3.14. Transient behavior of the Ćuk-converter.

3.5.5 Measurement and Simulation Results of the Closed-loop

This subsection is devoted to the comparative results of the measured and simulated closed-loop behavior with the non-linear state feedback H_2-controller with integral term from (3.121). The operating point of the duty ratio is fixed at $\bar{d}_A = 0.485$ and from this with $U_0 = 12$ V and $G_L = 1/22.36$ S, we get $[\bar{\imath}_{L,1}, \bar{\imath}_{L,2}, \bar{u}_{C,1}, \bar{u}_{C,2}] = [0.44, 22.01, -0.45, -10.0]$. The parameters of the controller (3.121) are chosen as $\beta = 0.001$ and $p_{22} = 0.2$ and a sampling time of $2 \cdot 10^{-4}$ s is used. Fig. 3.15 shows the simulated and measured output voltage $u_{C,2}$ and the corresponding duty ratio d_A for the reference input $u_{C,2,ref} = -10 + \Delta u_{C,2,ref}$ in V with

$$\Delta u_{C,2,ref} = 9\sigma \left(t - 2 \cdot 10^{-2}\right) - 19\sigma \left(t - 7 \cdot 10^{-2}\right) + 10\sigma \left(t - 12 \cdot 10^{-2}\right) . \tag{3.123}$$

Here $\sigma(t)$ denotes the unit step. Fig. 3.16 depicts the simulated and measured transient responses of the output voltage $u_{C,2}$ and of the corresponding duty ratio d_A, when the converter is subjected to a load variation $G_L = 1/22.36 +$

ΔG_L in S with

$$\Delta G_L = \left(\frac{1}{80} - \frac{1}{22.36}\right)\sigma\left(t - 2\cdot 10^{-2}\right) - \left(\frac{1}{80} - \frac{1}{8}\right)\sigma\left(t - 6\cdot 10^{-2}\right) .$$
(3.124)

It is easy to see that the proposed controller has excellent tracking as well as disturbance rejection behavior and that the duty ratio d_A remains within the admissible boundaries.

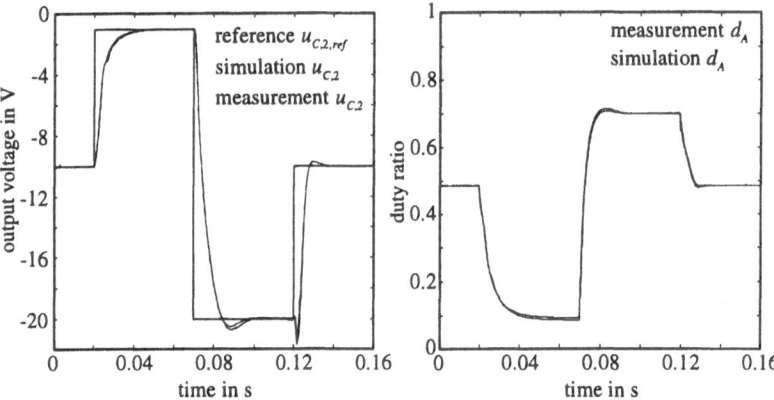

Fig. 3.15. Measurement and simulation results of the closed-loop system for the tracking behavior.

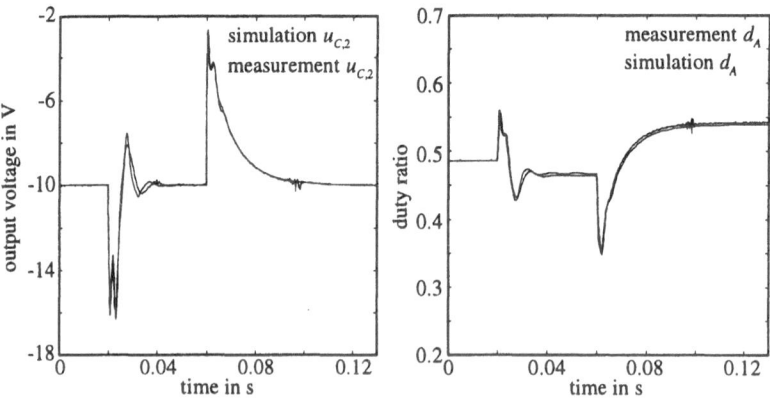

Fig. 3.16. Measurement and simulation results of the closed-loop system in the case of load variations.

3.6 Electromechanical System: The Energy/Co-energy Concept

In electromechanical systems there is an energy conversion process between the mechanical and the electrical part via the electric and magnetic coupling fields. Throughout this section we will assume that the electromechanical energy conversion itself is lossless and that the losses, such as ohmic losses or friction, can be separated from the energy storage mechanism and can be represented in the form of external loss terminals, like resistors or dampers etc. Let us assume that the electromechanical energy conversion occurs solely via the magnetic and electric coupling fields of the inductors and capacitors respectively.

Without restriction of generality and for the sake of clarity, we will consider electric networks where the set of inductor currents i_k, $k \in L$ and capacitor voltages u^l, $l \in C$ are linearly independent (see, Subsection 3.2.1). With slight modifications the whole theory can be extended to the case of dependent sets of inductor currents and capacitor voltages of Subsection 3.2.2. Remember that in Subsection 3.2.1 the states of the electric network are confined to a submanifold $\mathcal{N} = \{(i, u) \in \mathcal{N}_u \,|\, uD_I = 0, D_U^T i = 0\}$ of the space of unrestricted states $\mathcal{N}_u$, where the currents i and voltages u satisfy KCL and KVL. Let us consider that the mechanical part of the electromechanical system has n degrees of freedom and is represented by the generalized coordinates $x = (x_1, \dots, x_n) \in R^n$. Naturally, the generalized velocities $v = \frac{d}{dt}x$ are elements of the tangent space TR^n of R^n. Thus, the configuration space of the electromechanical system takes the form $\mathcal{N}_{em} = \mathcal{N} \times R^n \times TR^n$. Let the inductor currents i_k, $k \in L$, the capacitor voltages u^l, $l \in C$, the generalized coordinates x and the generalized velocities v together with the coordinate function $z : (i_j, u^j, x, v) \rightarrow (i_k, u^l, x, v)$, $j \in B$, $k \in L$, $l \in C$ be a chart of the configuration space $\mathcal{N}_{em}$. Further, M denotes the set of all inductors and capacitors where an energy conversion takes place and whose flux and charge linkages, ψ^j and q_j, depend on x. Subsequently, x_k, $k \in M$ means that the generalized coordinate x_k is linked with an inductor or capacitor. Then the differential equations of the dynamic elements inductor and capacitor due to (3.15) can be generalized in the form

$$
\begin{aligned}
\frac{d}{dt}\psi^j \left(\dots, i_l, \dots, x_k, \dots \right) &= u^j \quad \text{with} \quad j, l \in L, \quad k \in M \\
\frac{d}{dt}q_j \left(\dots, u^l, \dots, x_k, \dots \right) &= i_j \quad \text{with} \quad j, l \in C, \quad k \in M.
\end{aligned}
\tag{3.125}
$$

Analogous to (3.16) the energies $\hat{w}_L$ and $\hat{w}_C$ stored in the inductors and capacitors are given by

$$
\hat{w}_L = \int_\gamma \sum_{j \in L} i_j d_e \psi^j \quad \text{and} \quad \hat{w}_C = \int_\gamma \sum_{j \in C} u^j d_e q_j
\tag{3.126}
$$

with $\gamma(t)$ as a solution curve of the electromechanical system and d_e as the exterior derivative operating in the variables (i_l, u^j), $l \in L$, $j \in C$, i.e.

$$d_e f = \sum_{j \in L} \frac{\partial f}{\partial i_j} di_j + \sum_{j \in C} \frac{\partial f}{\partial u^j} du^j . \tag{3.127}$$

Let us also assume that the integrals in (3.126) are path independent, i.e. the relations

$$d_e \left(\sum_{j \in L} i_j d_e \psi^j \right) = 0 \quad \text{and} \quad d_e \left(\sum_{j \in C} u^j d_e q_j \right) = 0 \tag{3.128}$$

are satisfied.

Due to the assumption that the energy conversion is lossless the energy balance principle states that the increase in the energy stored in the inductors and capacitors $\hat{w} = \hat{w}_L + \hat{w}_C$ equals the energy input from the electrical sources minus the mechanical energy output. By taking F^k, $k \in M$, as the generalized coupling force associated with the generalized coordinate x_k, $k \in M$, we get the relation

$$z^* \left(d\hat{w} - \sum_{j \in \{L,C\}} i_j u^j dt + \sum_{k \in M} v_k F^k dt \right) = 0 \tag{3.129}$$

or equivalently

$$z^* \left(d\hat{w} - \sum_{j \in L} i_j d\psi^j - \sum_{j \in C} u^j dq_j + \sum_{k \in M} F^k dx_k \right) = 0 \tag{3.130}$$

with z^* as the pullback of the function $z : (i_j, u^j, x, v) \to (i_k, u^l, x, v)$, $j \in B$, $k \in L$, $l \in C$. Note that the exterior derivative d operates in all variables (i_k, u^l, x, v), $k \in L$, $l \in C$, i.e.,

$$df = \sum_{j \in L} \frac{\partial f}{\partial i_j} di_j + \sum_{j \in C} \frac{\partial f}{\partial u^j} du^j + \sum_{j=1}^{n} \frac{\partial f}{\partial x_j} dx_j + \sum_{j=1}^{n} \frac{\partial f}{\partial v_j} dv_j . \tag{3.131}$$

In order to determine an explicit expression of the generalized coupling forces F^k (see also [120]), we calculate in a first step $d\hat{w} = d\hat{w}_L + d\hat{w}_C$ with $\hat{w}_L$ and $\hat{w}_C$ from (3.126)

$$d\hat{w} = \sum_{j \in L} i_j d_e \psi^j + \sum_{j \in C} u^j d_e q_j +$$

$$\sum_{k \in M} \frac{\partial}{\partial x_k} \left(\int_\gamma \sum_{j \in L} i_j d_e \psi^j + \int_\gamma \sum_{j \in C} u^j d_e q_j \right) dx_k . \tag{3.132}$$

By applying Leibniz' rule we can rewrite (3.132) in the form

$$
d\hat{w} = \sum_{j \in L} i_j d_e \psi^j + \sum_{j \in C} u^j d_e q_j +
$$

$$
\sum_{k \in M} \frac{\partial}{\partial x_k} \left(\int_{\gamma} \sum_{j \in L} \left(d_e \left(i_j \psi^j \right) - \psi^j d i_j \right) \right) dx_k +
\tag{3.133}
$$

$$
\sum_{k \in M} \frac{\partial}{\partial x_k} \left(\int_{\gamma} \sum_{j \in C} \left(d_e \left(u^j q_j \right) - q_j d u^j \right) \right) dx_k
$$

or

$$
d\hat{w} = \sum_{j \in L} i_j d\psi^j + \sum_{j \in C} u^j d q_j -
$$

$$
\sum_{k \in M} \frac{\partial}{\partial x_k} \left(\int_{\gamma} \sum_{j \in L} \psi^j d i_j + \int_{\gamma} \sum_{j \in C} q_j d u^j \right) dx_k .
\tag{3.134}
$$

Substituting (3.134) into (3.130), we end up with

$$
\sum_{k \in M} z^* \left(-\frac{\partial}{\partial x_k} \left(\int_{\gamma} \sum_{j \in L} \psi^j d i_j + \int_{\gamma} \sum_{j \in C} q_j d u^j \right) + F^k \right) dx_k = 0 .
\tag{3.135}
$$

Since all dx_k are linearly independent, the expression in the bracket of (3.135) must vanish identically for all $k \in M$ and hence F^k takes the form

$$
F^k = \frac{\partial}{\partial x_k} z^* \left(\int_{\gamma} \sum_{j \in L} \psi^j d i_j + \int_{\gamma} \sum_{j \in C} q_j d u^j \right) .
\tag{3.136}
$$

For the sake of convenience we will henceforth drop the pullback operation z^*.

If the generalized coordinate x_k is associated only with either an inductor or a capacitor, the generalized coupling force F^k simplifies to

$$
F^k = \frac{\partial}{\partial x_k} \int_{\gamma} \sum_{j \in L} \psi^j d i_j \quad \text{or} \quad F^k = \frac{\partial}{\partial x_k} \int_{\gamma} \sum_{j \in C} q_j d u^j .
\tag{3.137}
$$

It can be immediately seen that the integrability conditions (3.128) imply

$$
d_e \left(\sum_{j \in L} \psi^j d i_j \right) = 0 \quad \text{and} \quad d_e \left(\sum_{j \in C} q_j d u^j \right) = 0 .
\tag{3.138}
$$

Hence we may define the functions

$$\breve{w}_L = \int_\gamma \sum_{j \in L} \psi^j \mathrm{d}i_j \quad \text{and} \quad \breve{w}_C = \int_\gamma \sum_{j \in C} q_j \mathrm{d}u^j \tag{3.139}$$

as the so-called co-energy functions of the inductors and capacitors respectively.

Remark 3.4. In the case of an electromechanical system with a linear magnetic or electrostatic characteristic the flux and charge linkage can be expressed in the form

$$\psi^j = \sum_{l \in L} \psi_x^{jl}(x) \, i_l \quad \text{or} \quad q_j = \sum_{l \in C} q_{jl,x}(x) \, u^l . \tag{3.140}$$

By means of the integrability conditions (3.128) or (3.138), we have $\psi_x^{jl}(x) = \psi_x^{lj}(x)$ and $q_{jl,x}(x) = q_{lj,x}(x)$ and thus, we may immediately conclude that in the linear case the expressions for the energy functions $\hat{w}_L$ and $\hat{w}_C$ from (3.126) and the co-energy functions $\breve{w}_L$ and $\breve{w}_C$ from (3.139) are equal.

3.6.1 Simple Application: Specific Influence on the Electrostatic/Electromagnetic Coupling Force

The energy conversion mechanism of electromechanical systems is often advantageously used in sensor and actuator applications, such as condenser microphones, capacitive acceleration sensors, electromagnetic valves, electrodynamic shakers etc. In general the coupling force F^k due to (3.137) is a non-linear function of the generalized coordinate x_k, but in many applications this non-linear dependence is undesirable.

Let us take as an example a silicon micro-machined electrostatic transducer with a moving membrane electrode and a rigid backplate electrode as shown in Fig. 3.17. The supply voltage applied between the two electrodes causes an electrostatic coupling force acting on the moving electrode. The designer of such an electrostatic transducer has to cope with two contradictory demands, namely the sensitivity should be as high as possible and the non-linear distortion factor as low as possible. An increase in sensitivity can be achieved by a reduced distance d_0 between the electrodes, a bigger effective electrode area $R_0^2\pi$ or by a higher supply voltage. But all these design changes cause the electrostatic coupling force to become a stronger influencing factor. Hence, according to the non-linear nature of the electrostatic force, the non-linear distortion factor is getting bigger. Furthermore, an increase in the electrostatic force level may even cause the membrane and the backplate electrode to stick. The reader should refer to e.g., [69] for a more detailed treatment of this topic.

Other examples are electromagnetic valves, where the air-gap configuration is often constructed in such a way that there is a linear (affine) stationary

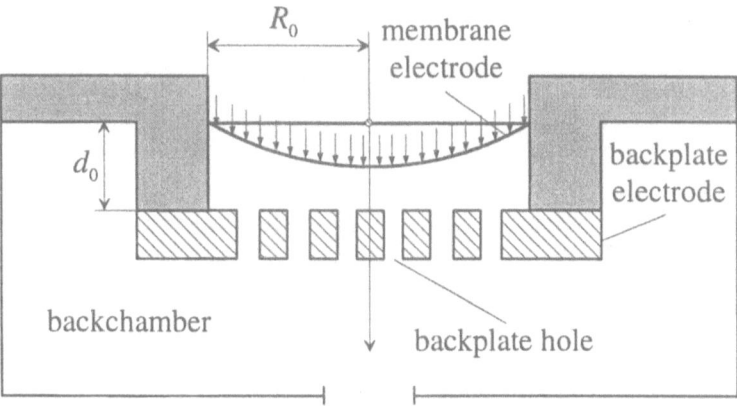

Fig. 3.17. Schematic diagram of a silicon micro-machined electrostatic transducer.

characteristic between the magnetic coupling force acting on the plunger and the piston position (see, e.g., [56]). Apart from the constructional influence on the system's behavior it is also possible to control the system so that the closed-loop meets the designer's requirements. In the latter case, the remaining design freedom may be even used to optimize the system from a point of view which cannot be addressed by control. In particular, the pleasing physical structure of the coupling mechanism of electromechanical systems offers a variety of methods for a physically based controller design. Let us take as an example the simple electromagnetic valve of Fig. 1.4 from Chapter 1 with $F_{ext} = 0$ and let us assume an ideal coil current controller such that the coil current i_L serves as a control input. Then the mathematical model of the electromagnetic valve of Fig. 1.4 written as a PCHD-system reads as

$$\frac{\mathrm{d}}{\mathrm{d}t}x = (J(x) - S(x))\left(\frac{\partial V}{\partial x}\right)^T + g(x)i_L^2 \tag{3.141}$$

with the state $x^T = [z, mv]$, the storage function

$$V(x) = \frac{1}{2}\left(cx_1^2 + \frac{1}{m}x_2^2\right) \tag{3.142}$$

and

$$J(x) = \begin{bmatrix} 0 & 1 \\ -1 & 0 \end{bmatrix}, \ S(x) = \begin{bmatrix} 0 & 0 \\ 0 & d \end{bmatrix}, \ g(x) = \begin{bmatrix} 0 \\ \dfrac{1}{2}\dfrac{\partial L(x_1)}{\partial x_1} \end{bmatrix}, \tag{3.143}$$

where $L(x_1)$ is the inductance of the magnetic circuit due to (1.25)

$$L(x_1) = \frac{\mu_0 N^2 D^2 \pi (D+\delta)\pi b}{4(h-x_1)(D+\delta)\pi b + \delta D^2 \pi}. \tag{3.144}$$

If we want the electromagnetic valve to behave like a linear spring-mass-damper system, with a spring constant $\tilde{c}$ and an initial position $x_d^T = [x_{1,d}, 0]$, then the storage function of the closed-loop has to take the form

$$V_c = \frac{1}{2}\left(\tilde{c}\,(x_1 - x_{1,d})^2 + \frac{1}{m}x_2^2\right). \tag{3.145}$$

Clearly, the storage function of the closed-loop V_c has a strict minimum at x_d and $V_c(x) - V_c(x_d)$ is positive definite. Following the passivity-based control concept of Section 2.4 the control law must satisfy the relation (see (2.60))

$$g(x)\,i_L^2 = (J(x) - S(x))\left(\frac{\partial V_a}{\partial x}\right)^T \tag{3.146}$$

with $V_a = V_c - V$. A short calculation leads to the control law

$$i_L = \sqrt{\frac{2\,(cx_1 - \tilde{c}\,(x_1 - x_{1,d}))}{\partial L\,(x_1)\,/\partial x_1}}, \tag{3.147}$$

where $\tilde{c}$ and $x_{1,d}$ must be chosen such that the expression under the square root is always positive.

By means of the so-called damping injection method we can also easily change the damping coefficient of the closed-loop linear spring-mass-damper system from d to $\tilde{d}$. Following (2.62) and (2.64) from Section 2.4 with $S_d(x) = \left(\tilde{d} - d\right)\left(\frac{1}{2}\frac{\partial L(x_1)}{\partial x_1}\right)^{-2}$, we immediately get the modified control law

$$i_L = \sqrt{\frac{2\left(cx_1 - \tilde{c}\,(x_1 - x_{1,d}) + \left(d - \tilde{d}\right)\frac{\mathrm{d}}{\mathrm{d}t}x_1\right)}{\partial L\,(x_1)\,/\partial x_1}}. \tag{3.148}$$

Remark 3.5. This method is not restricted to electromechanical systems with a linear magnetic or electrostatic characteristic. We can replace the left hand side of (3.146) by $\frac{\partial}{\partial x_k}\tilde{w}_L$ with $\tilde{w}_L$ as the co-energy function due to (3.139). The only difference is that in general we no longer get an explicit expression for the control law as it is the case in (3.147) or (3.148). But if $\left(\frac{\partial}{\partial i_L}\left(\frac{\partial}{\partial x_k}\tilde{w}_L\right)\right)(x_k, i_L) \neq 0$, we may apply the implicit function theorem (e.g., [59]) and we locally get a unique solution for i_L.

To summarize, we see that by means of control, we can specifically influence the dynamic and stationary behavior of electromechanical sensors and actuators and in this way we are able to improve their quality.

4. Mechanical PCH-systems

This chapter deals with finite- and infinite-dimensional mechanical systems which have the representation of a PCH-(port-controlled Hamiltonian) system as defined in Section 1.4. Although this class of mechanical systems contains no dissipative forces, as they are always present in the form of natural damping in every realistic application, the controllers designed on the basis of these systems are of practical importance. The reason is that the undamped model is something like the "worst case according to damping" for a controller design which is essentially based on damping injection. In many mechanical systems with weak damping, as is the case for flexible structures, the resulting controllers show sufficient robustness against parameter inaccuracies, in particular concerning the damping behavior.

In view of the application of this chapter, namely control of infinite-dimensional smart beam structures, we are interested in a mathematical framework for a uniform description of the finite- and infinite-dimensional case. In literature, symplectic or Poisson manifolds serve as a natural geometric oriented formulation of Hamiltonian systems, see, e.g., [1], [16], [18], [93], [106]. In this chapter, we will briefly summarize some essential results of the Poisson bracket approach which are important for the control applications being considered. We will also assume that the PCH-systems considered are described in a suitable vector space with a well-defined Poisson bracket. On this basis we will formulate the non-linear H_2-design (see Section 2.1), the non-linear H_∞-design (see Section 2.3), the PD-(proportional differential) controller design and the idea of disturbance compensation for finite- and infinite-dimensional PCH-systems. The stability investigations of infinite-dimensional systems is much more complicated than in the finite-dimensional case because the compactness of the level sets of the Lyapunov functions are no longer automatically ensured. We will not address this problem here, but the reader is referred to e.g., [1], [18], [92] and the references cited there for more details.

By means of an infinite-dimensional piezoelectric composite beam structure we will apply the different control strategies developed so far for PCH-systems. The feasibility of these control concepts essentially relies on the fact that the piezoelectric structures allow a spatial distribution of the piezoelectric sensor and actuator layers. The design of the spatial pattern of the sensor and actuator electrodes is an additional degree of freedom and can be regarded as a part of the control synthesis task. In this way, we are able to design the sensor and actuator layers in such a way that they are collocated and hence the well known effects of observation/actuation spillover can be prevented.

4.1 Fundamentals of Lagrangian and Hamiltonian Systems

4.1.1 The Finite-dimensional Case

Let us consider a mechanical system with no external and dissipative forces which can be locally described by the coordinates $x = (x_1, \dots, x_n)$ of the configuration manifold $\mathcal{M}$. The Lagrangian $L(x, v)$ can be regarded as a function from the tangent bundle $T\mathcal{M}$ of $\mathcal{M}$ to R with the system velocities $v_i = \frac{d}{dt} x_i$, $i = 1, \dots, n$. Usually, the Lagrangian is the kinetic minus the potential energy, but this is not always the case, e.g., if we calculate Lagrange's equations with electromagnetism, see, e.g., [32]. From Hamilton's variational principle, which states that the action integral $\int_{t_1}^{t_2} L(x, v)\, dt$ is extremized for curves $x(t)$ connecting two fixed endpoints $x(t_1)$ and $x(t_2)$, we can derive the well-known Euler-Lagrange equations [36], [93]

$$\frac{d}{dt} \frac{\partial L}{\partial v_i} - \frac{\partial L}{\partial x_i} = 0 \quad \text{for} \quad i = 1, \dots, n \ . \tag{4.1}$$

Now, by means of the Legendre transform $(x, v) \rightarrow (x, p)$ with the conjugate momenta $p^i = \frac{\partial L}{\partial v_i}$, $i = 1, \dots, n$, the Euler-Lagrange equations (4.1) can be transformed into the equivalent Hamilton's equations

$$\frac{d}{dt} \begin{bmatrix} x \\ p \end{bmatrix} = \underbrace{\begin{bmatrix} 0 & I \\ -I & 0 \end{bmatrix}}_{J} \begin{bmatrix} \frac{\partial H}{\partial x} \\ \frac{\partial H}{\partial p} \end{bmatrix} \tag{4.2}$$

with the identity matrix I and the associated Hamiltonian function or energy function

$$H(p, x) = \sum_{i=1}^{n} v_i p^i - L(x, v) \ . \tag{4.3}$$

In order to express v in (4.3) as a function of p and x, we assume that the transformation $(x, v) \rightarrow (x, p)$ is invertible. Obviously, if the matrix $\left[\frac{\partial^2}{\partial v_i \partial v_j} L \right]$ is regular the implicit function theorem guarantees that the transformation is locally invertible. In this case, the Lagrangian L is said to be regular or non-degenerate, see, e.g., [93]. In more general terms, the Legendre transform is a fiber derivative which defines a map from the tangent bundle $\mathcal{T}\mathcal{M}$ to the cotangent bundle $\mathcal{T}^*\mathcal{M}$ of the configuration manifold $\mathcal{M}$ [1], [93]. It is easy to see that (4.2) is a PCH-system as shown in Section 1.4.

Remark 4.1. If we assume a mechanical system with a Lagrangian of the form $L(x, v) = w_k(x, v) - w_p(x)$, with the potential energy w_p and the kinetic energy $w_k = \frac{1}{2}v^T M(x) v$, where $M(x)$ denotes the positive definite generalized inertia matrix, then it is quite clear that the energy function H of (4.3) is equivalent to the sum of kinetic and potential energy.

Remark 4.2. Let us consider an electromechanical system as presented in Section 3.6 where the inductor currents i_k, $k \in L$, the capacitor voltages u^l, $l \in C$, the generalized coordinates x and the generalized velocities v together with the coordinate function $z : (i_j, u^j, x, v) \rightarrow (i_k, u^l, x, v)$, $j \in B$, $k \in L$, $l \in C$ form a chart of the configuration space. Remember that with i_j, u^j, $j \in B$ we mean the currents and voltages of all branches of the electric network. Further, let x_k, $k \in M$ denote those generalized coordinates x which are linked with an inductor or capacitor. The electrical part of the system equations of an electromechanical system is given by (3.125)

$$
\begin{aligned}
\frac{d}{dt}\psi^j(\ldots, i_l, \ldots, x_k, \ldots) &= u^j \quad, \quad j, l \in L, \quad k \in M \\
\frac{d}{dt}q_j(\ldots, u^l, \ldots, x_k, \ldots) &= i_j \quad, \quad j, l \in C, \quad k \in M,
\end{aligned}
\tag{4.4}
$$

with the flux and charge linkages ψ^j and q_j. Provided that the existence of the co-energy functions $\breve{w}_L$ and $\breve{w}_C$ for the inductors and capacitors from (3.139) can be ensured, the mechanical part of the equations of motion can be easily derived by means of the Euler-Lagrange equations (4.1). We just have to extend the Lagrangian L in the form $\breve{L} = L + \breve{w}_L + \breve{w}_C$ with

$$
\breve{w}_L = z^* \left(\int_\gamma \sum_{j \in L} \psi^j \, di_j \right) \quad \text{and} \quad \breve{w}_C = z^* \left(\int_\gamma \sum_{j \in C} q_j \, du^j \right)
\tag{4.5}
$$

where z^* denotes the pullback of the coordinate function z.

In the literature, symplectic or Poisson manifolds are used for a natural geometric oriented formulation of Hamiltonian systems [1], [16], [18], [93], [106]. Here, we will just briefly summarize some essential results of the Poisson bracket approach which are important for the subsequent considerations. We are particularly interested in a mathematical framework for a uniform description of finite- and infinite-dimensional Hamiltonian systems, see also [125].

A Poisson manifold $\mathcal{P}$ is a manifold with a Poisson bracket $\{\cdot, \cdot\}$. The Poisson bracket assigns to two smooth real-valued functions F, $G : \mathcal{P} \rightarrow$

R another smooth real-valued function $\{F, G\}$ which is bilinear and skew-symmetric. Furthermore, the Poisson bracket satisfies the Jacobi-identity

$$\{\{F, G\}, H\} = \{\{F, H\}, G\} + \{F, \{G, H\}\} \tag{4.6}$$

and Leibniz' rule

$$\{F, HG\} = \{F, H\} G + H \{F, G\} \tag{4.7}$$

with the smooth real-valued functions F, H, G.

Let $\mathcal{P}$ be a Poisson manifold with a smooth function $H : \mathcal{P} \to R$. Then there exists a unique vector field $X_H \in T\mathcal{P}$, with $T\mathcal{P}$ as the tangent bundle of $\mathcal{P}$, such that the relation

$$L_{X_H} F = \{F, H\} \tag{4.8}$$

holds for all smooth functions $F : \mathcal{P} \to R$ with $L_{X_H} F$ as the Lie-derivative of F along X_H, see, e.g., [93], [106]. The vector field X_H is then called a Hamiltonian vector field associated with the Hamiltonian function H. Consequently, the rate of change of a function F along the integral curve of X_H is given by

$$\frac{\mathrm{d}}{\mathrm{d}t} F = \{F, H\} . \tag{4.9}$$

Equation (4.9) is often called the Poisson bracket form of the equations of motion [93]. Now, it is easy to classify a constant of motion, namely a function F on $\mathcal{P}$ is a constant of motion if and only if $\{F, H\} = 0$. Thus, due to the skew-symmetry of the Poisson bracket it follows directly that the Hamiltonian function H is a constant of motion. In other words the total energy H is conserved along the integral curves of the system.

Let us consider a Poisson manifold $\mathcal{P}$ with local coordinates (t, u^β), $\beta = 1, \dots, s$, where t denotes the independent variable and u represents the dependent variables. In the following we will assume that the functions F and H on $\mathcal{P}$ do not depend explicitly on t, but this is no restriction of generality, see, e.g., [106]. Thus, in the local coordinates u the Poisson bracket reads as

$$\{F, H\} = \sum_{\beta=1}^{s} \sum_{\gamma=1}^{s} J^{\beta\gamma}(u) \frac{\partial F}{\partial u^\beta} \frac{\partial H}{\partial u^\gamma} \tag{4.10}$$

with $J^{\beta\gamma}(u) = \{u^\beta, u^\gamma\}$ as the so-called structure functions of the Poisson manifold $\mathcal{P}$. To show the relation (4.10), at first, we write the Hamiltonian vector field X_H in the local coordinates u in the form $X_H = \sum_{\beta=1}^{s} a^\beta(u) \frac{\partial}{\partial u^\beta}$ with the coordinate functions $a^\beta(u)$, see [106]. Then by (4.8) we have

$$\{u^\beta, H\} = a^\beta(u) = -\{H, u^\beta\} \tag{4.11}$$

and hence $\{F, H\}$ can be written in the form

$$\{F, H\} = \sum_{\beta=1}^{s} a^\beta(u) \frac{\partial F}{\partial u^\beta} = -\sum_{\beta=1}^{s} \{H, u^\beta\} \frac{\partial F}{\partial u^\beta}. \tag{4.12}$$

Replacing F by H and H by u^β in (4.12), we get the relation

$$\{H, u^\beta\} = -\sum_{\gamma=1}^{s} \{u^\beta, u^\gamma\} \frac{\partial H}{\partial u^\gamma} \tag{4.13}$$

and this leads us directly to the result

$$\{F, H\} = \sum_{\beta=1}^{s} \sum_{\gamma=1}^{s} \{u^\beta, u^\gamma\} \frac{\partial F}{\partial u^\beta} \frac{\partial H}{\partial u^\gamma}. \tag{4.14}$$

Thus, in local coordinates u the Hamiltonian vector field takes the form

$$X_H = \sum_{\beta=1}^{s} \sum_{\gamma=1}^{s} J^{\beta\gamma}(u) \frac{\partial H}{\partial u^\gamma} \frac{\partial}{\partial u^\beta} \tag{4.15}$$

and the Hamilton's equations read as

$$\frac{d}{dt} u^\beta = \{u^\beta, H\} = \sum_{\gamma=1}^{s} J^{\beta\gamma}(u) \frac{\partial H}{\partial u^\gamma}. \tag{4.16}$$

Moreover, if $\mathcal{P}$ is a Poisson manifold then there exists a unique linear map $\Omega^\# : T^*\mathcal{P} \to T\mathcal{P}$, coming from a bi-vector, such that the relation

$$\frac{d}{dt} u = \Omega^\#(dH(u)) = X_H(u) \tag{4.17}$$

holds for all smooth functions H with $u \in \mathcal{P}$, $dH \in T^*\mathcal{P}$, the cotangent bundle of $\mathcal{P}$, and $X_H \in T\mathcal{P}$, the tangent bundle of $\mathcal{P}$. The rank of the Poisson manifold at u is then defined by the rank of the linear map $\Omega^\#$ at u, see, e.g., [106]. In local coordinates u the rank of the Poisson manifold $\mathcal{P}$ equals the rank of the so-called structure matrix $J(u) = [J^{\beta\gamma}(u)]$. Clearly, $J(u)$ is a skew-symmetric matrix. Now, by Darboux' Theorem (for bi-vectors) we can always find local canonical coordinates $u^\alpha = (x_i, p^i)$, $\alpha = 1, \ldots, s$, $i = 1, \ldots, n$, $s = 2n$, for a Poisson manifold $\mathcal{P}$ with constant rank such that the Poisson bracket takes the form

$$\{F, G\} = \sum_{i=1}^{n} \left(\frac{\partial F}{\partial x_i} \frac{\partial G}{\partial p^i} - \frac{\partial F}{\partial p^i} \frac{\partial G}{\partial x_i} \right). \tag{4.18}$$

From (4.15) we see also that the Hamiltonian vector field X_H in canonical coordinates $u = (x, p)$ is given by

$$X_H = \sum_{i=1}^{n} \left(-\frac{\partial H}{\partial x_i} \frac{\partial}{\partial p^i} + \frac{\partial H}{\partial p^i} \frac{\partial}{\partial x_i} \right) . \tag{4.19}$$

Clearly, $\left(x_i(t), p^i(t) \right)$ is an integral curve of X_H if and only if Hamilton's equations (4.16) hold. Moreover, the structure matrix $J(u)$ in canonical coordinates (x, p) has the simple form

$$J = \begin{bmatrix} 0 & I \\ -I & 0 \end{bmatrix} \tag{4.20}$$

with the identity matrix I.

Remark 4.3. A Poisson manifold $\mathcal{P}$ is called symplectic if the associated Poisson bracket is nondegenerate i.e., the Poisson manifold $\mathcal{P}$ has maximal rank everywhere, see, e.g., [18], [93], [106] for more details.

4.1.2 The Infinite-dimensional Case

To begin with it should be emphasized that an exact mathematical formulation of the infinite-dimensional case requires many difficult technicalities. This is beyond the scope of this work. For details the interested reader is referred to [1], [18] and in particular [106]. Here, we intend only to point out the parallels to the finite-dimensional case. One of the main problems with evolutionary equations is that Darboux' Theorem is no longer valid. Nevertheless, the Poisson bracket approach of the previous subsection can be extended to infinite-dimensional Hamiltonian systems.

Firstly, let us introduce some useful notation. Consider a $(1 + r + s)$-dimensional smooth space $\mathcal{P} = R \times \mathcal{D} \times \mathcal{M}$ with local coordinates $\left(t, z^\alpha, u^\beta \right)$, $\alpha = 1, \dots, r$ and $\beta = 1, \dots, s$. Here, the time t and the spatial coordinates z denote the independent variables and u is used to denote the dependent variables. The space $\mathcal{P}^{(n)} = R \times \mathcal{D} \times \mathcal{M}^{(n)}$, whose coordinates represent the independent variables, the dependent variables and all derivatives of the dependent variables up to the order n is called the n-th order jet space of $\mathcal{P}$, see [106]. A smooth real-valued function $f\left(t, z^1, \dots, z^r \right)$ of $(r + 1)$ independent variables has $\binom{r+k}{k}$ different k-th order partial derivatives. Therefore, in order to make the notation clearer, it makes sense to use the symmetric multi-index notation

$$\frac{\partial^k}{\partial z^{j_1} \partial z^{j_2} \dots \partial z^{j_k}} f = f_J, \quad f_0 = f \tag{4.21}$$

with $z^0 = t$. In this notation $J = (j_1, \dots, j_k)$ is an unordered symmetric multi-index, with $0 \le j_i \le r$, $i = 1, \dots, k$ indicating the derivatives being taken and $\#J = k$ is the order of J. In the following $u^{(n)}$ always refers to an element of $\mathcal{M}^{(n)}$ and hence its components are of the type u_J^β, with $\beta = 1, \ldots, s$ and J running over all symmetric multi-indices (4.21) for $0 \leq k \leq n$. Henceforth, in order to simplify the notation, we will write $X(f)$ instead of $L_X f$ for the Lie derivative of a functional f along a vector field X.

In the infinite-dimensional case the Hamiltonian function has to be replaced by a functional of the form

$$H = \int_D h\left(z, u^{(n)}\right) \omega_v, \quad \omega_v = dz^1 \wedge \ldots \wedge dz^r. \tag{4.22}$$

Let us assume that the density h is well defined on $\mathcal{D}$ for all times $t \geq 0$. Further, we have to add suitable conditions for $u^{(n)}$ on the boundary $\partial \mathcal{D}$ of $\mathcal{D}$. For the sake of simplicity the boundary conditions will be specified only in the applications, because they are not relevant for the following considerations. Let φ_τ be a one parameter group $\left(\tilde{t}, \tilde{z}, \tilde{u}^{(n)}\right)(\tau) = \varphi_\tau\left(t, z, u^{(n)}\right)$ which acts on $\mathcal{P}^{(n)}$ such that the relations $\tilde{t}(\tau) = t$ and $\tilde{z}(\tau) = z$ are met. Then the infinitesimal generator $\hat{X} \in T\left(\mathcal{P}^{(n)}\right)$ of φ_τ takes the form $\hat{X} = \mathrm{pr}^{(n)} X$ with the evolutionary field $X = \sum_{\beta=1}^s a^\beta\left(t, z, u^{(n)}\right) \frac{\partial}{\partial u^\beta}$, $X \in T(\mathcal{P})$. Here $\mathrm{pr}^{(n)} X$ denotes the n-th prolongation of X given by [106]

$$\mathrm{pr}^{(n)} X = \sum_{\beta=1}^s \sum_J D_J a^\beta \frac{\partial}{\partial u_J^\beta}, \quad D_J = \frac{d}{dz^{j_1}} \cdots \frac{d}{dz^{j_k}} \tag{4.23}$$

with the symmetric multi-index $J = (j_1, \ldots, j_k)$ and the total derivatives

$$\frac{d}{dz^i} = \frac{\partial}{\partial z^i} + \sum_{\beta=1}^s \sum_J \frac{\partial}{\partial z^i} u_J^\beta \frac{\partial}{\partial u_J^\beta}, \quad i = 0, \ldots, r. \tag{4.24}$$

The sum $\sum_J$ in (4.23) and (4.24) is taken over all symmetric multi-indices J up to the order n, i.e. $0 \leq \#J \leq n$. The functional (4.22) is invariant with respect to φ_τ, if and only if the relation

$$\mathrm{pr}^{(n)} X(H) = \int_D \mathrm{pr}^{(n)} X(h) \, \omega_v = \int_D \sum_{\beta=1}^s a^\beta E_\beta(h) \, \omega_v = 0 \tag{4.25}$$

is met, with the Euler operators

$$E_\beta = \sum_J (-1)^{\#J} D_J \frac{\partial}{\partial u_J^\beta}, \quad \beta = 1, \ldots, s. \tag{4.26}$$

The third term of (4.25) is obtained by simple integration by parts and the fact that due to the imposed boundary conditions the terms on the boundary $\partial \mathcal{D}$ vanish.

Let J denote a skew-adjoint differential operator, then the choice

$$a^\beta = \sum_{\gamma=1}^{s} J^{\beta\gamma} \left(\mathsf{E}_\gamma \left(h \right) \right) \tag{4.27}$$

guarantees that (4.25) holds. Recall that a differential operator J is called skew-adjoint if the following relation

$$\int_{\mathcal{D}} \sum_{\beta=1}^{n} \sum_{\gamma=1}^{m} \left(\xi^\beta J^{\beta\gamma} \left(\chi^\gamma \right) - \chi^\gamma J^{\gamma\beta} \left(\xi^\beta \right) \right) \omega_v = 0 \tag{4.28}$$

is fulfilled. Now, we are ready to define the bracket operation

$$\{F,H\} = \int_{\mathcal{D}} \sum_{\beta=1}^{s} \sum_{\gamma=1}^{s} \mathsf{E}_\beta \left(f \right) J^{\beta\gamma} \left(\mathsf{E}_\gamma \left(h \right) \right) \omega_v \tag{4.29}$$

for the functionals $F = \int_{\mathcal{D}} f \left(z, u^{(n)} \right) \omega_v$ and $H = \int_{\mathcal{D}} h \left(z, u^{(n)} \right) \omega_v$. Obviously, the bracket is bilinear and skew-symmetric by its definition. Furthermore, if and only if (4.29) satisfies the Jacobi-identity (4.6) then the bracket defined by (4.29) corresponds with the Poisson bracket, see, e.g., [106]. In this case the differential operator J is called Hamiltonian. The Leibniz rule (4.7) has no counterpart for infinite-dimensional Hamiltonian systems since there exists no well defined multiplication of functionals.

Let J be a Hamiltonian operator with a Poisson bracket as defined in (4.29). Analogous to (4.8) there exists a unique vector field X_H to the functional $H = \int_{\mathcal{D}} h \left(z, u^{(n)} \right) \omega_v$ such that the relation

$$\frac{\mathrm{d}}{\mathrm{d}t} F = \mathrm{pr}^{(n)} X_H \left(F \right) = \{F,H\} \tag{4.30}$$

is satisfied for all functionals $F = \int_{\mathcal{D}} f \left(z, u^{(n)} \right) \omega_v$, see [106]. The vector field X_H, given by

$$X_H = \sum_{\beta=1}^{s} \sum_{\gamma=1}^{s} J^{\beta\gamma} \left(\mathsf{E}_\gamma \left(h \right) \right) \frac{\partial}{\partial u^\beta}, \tag{4.31}$$

is then called Hamiltonian vector field associated with the Hamiltonian functional $H = \int_{\mathcal{D}} h \left(z, u^{(n)} \right) \omega_v$. Accordingly, the Hamilton's equations in the infinite-dimensional case read as

$$\frac{\partial}{\partial t} u^\beta = \{u^\beta,H\} = \mathrm{pr}^{(n)} X_H \left(u^\beta \right) = \sum_{\gamma=1}^{s} J^{\beta\gamma} \left(\mathsf{E}_\gamma \left(h \right) \right) , \quad \beta = 1, \ldots, s . \tag{4.32}$$

4.2 Controller Design Strategies

4.2.1 Preliminaries

For the controller design it is necessary to generalize the Hamilton's equations of (4.16), (4.32) for mechanical systems with external forces. Generally, these external forces represent the action of control, disturbance, dissipation or the interaction to the system environment. Here, the Lagrange-d'Alembert principle offers an effective way to describe the motion of a mechanical system subject to an external force field. This also optimally fits the geometric formulation of the Euler-Lagrange equations [93]. For the controller design of generalized finite-dimensional Euler-Lagrange systems the reader is encouraged to consult a very recent and important book [107].

From now on, we will focus on finite- and infinite-dimensional PCH-systems with external inputs e_j, $j = 1, \ldots, m'$, consisting of control inputs and disturbances. In the finite-dimensional case the associated Hamiltonian function reads as

$$H = H_0 - \sum_{j=1}^{m'} H_j e_j \, , \qquad (4.33)$$

where H_0 denotes the Hamiltonian function of the free system and H_j, $j = 1, \ldots, m'$ are the so-called interaction Hamiltonian functions, see, e.g., [103], [143]. For infinite-dimensional PCH-systems with external inputs the Hamiltonian function (4.33) has to be replaced by a functional H, with the Hamiltonian functional of the free system $H_0 = \int_D h_0 \left(z, u^{(n)} \right) \omega_v$ (see (4.22)) and the interaction Hamiltonian functionals $H_j = \int_D h_j \left(z, u^{(n)} \right) \omega_v$, $j = 1, \ldots, m'$. Here, we always assume that the external inputs e_j, $j = 1, \ldots, m'$ act in such a way on the system that the Hamiltonian functional has a representation of (4.33). Of course, this formulation also contains situations where the external inputs appear in the boundary conditions. Inserting H of (4.33) or its associated density h into (4.15) or (4.31), respectively, we can see that the external inputs e_j appear affine in the Hamiltonian vector field X_H, i.e. X_H has the form

$$X_H = X_{H_0} - \sum_{j=1}^{m'} e_j X_{H_j}. \qquad (4.34)$$

Analogous to the Poisson bracket form of the equations of motion (4.9) and (4.30) the change of a function F due to the motion of the PCH-system with the external inputs e_j is given by

$$\frac{\mathrm{d}}{\mathrm{d}t} F = \{F, H_0\} - \sum_{j=1}^{m'} \{F, H_j\} e_j \, . \qquad (4.35)$$

The special choice $y_j = H_j$, $j = 1, \ldots, m'$ for the output functions of a PCH-system is called the natural output [103]. The importance of this choice can be seen by calculating the time derivative of the Hamiltonian function (functional) H_0 of the free system along a solution of the PCH-system

$$\frac{dH_0}{dt} = \{H_0, H\} = \sum_{j=1}^{m'} \{H_j, H\} \, e_j = \sum_{j=1}^{m'} \frac{dy_j}{dt} e_j \, . \tag{4.36}$$

Equation (4.36) is exactly what we know as the energy balance equation and it states that the change of the internal (stored) energy H_0 of the PCH-system equals the flow of power into the system caused by the external inputs e_j. This configuration of inputs and outputs is also called the case of collocated sensors and actuators (e.g., [103]). For example, an attitude sensor is located at the same point as a torque actuator, or a displacement sensor is collocated with the corresponding compatible force actuator. A perfect sensor/actuator collocation has the big advantage that it usually provides a stable performance in the closed-loop feedback control, provided that the free system, i.e. no external inputs, is stable (e.g., [55]). In the case of infinite-dimensional mechanical systems with distributed sensors and actuators, such as smart structures with piezoelectric sensor and actuator layers, one can achieve a collocated sensor/actuator pairing by the additional degree of freedom of spatial shaping of the sensors and actuators (see, e.g., [50], [51], [68], [76], [118], [123]). Here, the design of the sensors and actuators becomes a part of the controller design.

Remark 4.4. Although dissipative forces, like e.g. the natural damping of flexible structures, are disregarded within the considered class of mechanical systems, the controllers designed on the basis of these systems are of practical importance. The reason is that the undamped model is something like the "worst case according to damping" for a controller design which is essentially based on damping injection. The resulting controllers show sufficient robustness against parameter inaccuracies of real-world applications, in particular concerning the damping behavior. At this point it is worth mentioning that in the literature a variety of models (Kelvin-Voigt, viscous, structural damping etc.) is available to add damping to flexible structures (see, e.g., [9] and the references therein). Generally, the damping parameters for a specific application can only be determined by means of experiments. In particular for non-linear systems this is rather a difficult task.

Next, we will present some essential results for the control of finite- and infinite-dimensional PCH-systems. It is noticeable that these design strategies do not depend on the specific structure of the Hamilton's equations but they require only the collocation of the sensors and actuators. Therefore, we do not need to distinguish between linear and non-linear or finite- and infinite-dimensional PCH-systems. Before starting with the controller design, we will briefly comment on the stability of infinite-dimensional systems.

4.2.2 Some Remarks Concerning the Stability of Infinite-dimensional Systems

A well-known method to investigate the stability of an equilibrium is given by Lyapunov's theory. But in the case of infinite-dimensional systems some additional aspects have to be taken into account, see, e.g., [1]. For finite-dimensional systems the compactness of the level sets of the Lyapunov function is automatically met [1]. The proof of this part in the infinite-dimensional case can be rather delicate. From the literature it is known that Lyapunov's stability is directly involved with the energy criterion for infinite-dimensional systems and with the so-called potential-well hypothesis. For this the reader is referred to [92] where interesting applications in the field of non-linear elasticity can also be found. Henceforth, our stability investigations of the infinite-dimensional PCH-systems under consideration are based on the following stability hypothesis: Let the Hamiltonian functional of the free system H_0 be a positive definite functional. Then, we assume that the condition $\frac{d}{dt}H_0 = \{H_0, H\} \leq 0$ implies the stability of the infinite-dimensional PCH-system.

Another peculiarity of infinite-dimensional systems, not known from the finite-dimensional case, is the so-called observation/actuation spillover [7]. This spillover effect may occur if the mathematical model for the controller design is based on a finite approximation of the infinite-dimensional system. The problem is that the control input can cause an unintentional excitation of the truncated modes and *vice versa* the truncated modes may have an undesired contribution to the sensor output. In both cases the performance of the closed-loop can be degraded, or in the worst case the system can even be destabilized. Similar effects can be observed by sensors/actuators, which are located at discrete points only and thus cannot sense/actuate those modes having a node at these points. However, this is why we will direct our attention to control strategies which *a priori* prevent spillover effects. The subsequent control strategies are applicable to both, finite- and infinite-dimensional systems.

4.2.3 Non-linear H_2-design for PCH-systems

Let us consider a PCH-system with the Hamiltonian function

$$H = H_0 - \sum_{j=1}^{m} H_j u_{C,j}, \tag{4.37}$$

where H_0 is the Hamiltonian function of the free system and $u_{C,j}$, $j = 1, \dots, m$ are the control inputs. Further, let us assume a perfect sensor/actuator collocation with the output functions

$$y_j = H_j, \quad j = 1, \dots, m . \tag{4.38}$$

Then the non-linear H_2-control problem from Section 2.1 is to find a control law $u_{C,j}$, $j = 1, \ldots, m$ such that the objective function

$$J_2 = \inf_{u_C \in L_2^m[0,\infty)} \frac{1}{2} \int_0^\infty \left(\left\| \frac{\mathrm{d}}{\mathrm{d}t} y \right\|^2 + \|u_C\|^2 \right) \mathrm{d}t \tag{4.39}$$

with $\| \ \|$ as the square norm is minimized with respect to u_C.

Proposition 4.1. *Given a PCH-system with the Hamiltonian function (4.37) and the natural outputs (4.38). Let us assume that $\{H_j, H_i\} = 0$ holds for all $i, j = 1, \ldots, m$ and the zero-state observability condition of Theorem 2.1 is satisfied. Then the control law*

$$u_{C,j} = -\frac{\mathrm{d}}{\mathrm{d}t} y_j, \quad j = 1, \ldots, m \tag{4.40}$$

solves the optimal non-linear H_2-control problem with respect to the objective function (4.39) [123].

Proof. For infinite-dimensional PCH-systems recall the stability hypothesis of Subsection 4.2.2. Following Section 2.1, the non-linear H_2-control problem is solved by finding a positive definite solution V of the associated HJBi (2.11)

$$\inf_{u_C \in L_2^m[0,\infty)} \left(\frac{\mathrm{d}}{\mathrm{d}t} V + \frac{1}{2} \left(\left\| \frac{\mathrm{d}}{\mathrm{d}t} y \right\|^2 + \|u_C\|^2 \right) \right) \leq 0 \tag{4.41}$$

or of the equivalent Poisson bracket representation

$$\inf_{u_C \in L_2^m[0,\infty)} \left(\{V, H_0\} + \sum_{j=1}^m \left(\frac{1}{2} \left(\{H_j, H_0\}^2 + u_{C,j}^2 \right) - \{V, H_j\} u_{C,j} \right) \right)$$
$$\leq 0. \tag{4.42}$$

From (4.42) one can immediately see that the optimal choice $u_{C,j}^*$ for $u_{C,j}$, $j = 1, \ldots, m$ is given by (compare with (2.9))

$$u_{C,j}^* = \{V, H_j\}. \tag{4.43}$$

The energy function of the free system H_0 serves as a suitable candidate for solving (4.42), namely $V = \rho H_0$, $\rho > 0$. Inserting V and $u_{C,j}^*$ into (4.42), we get the inequality

$$\frac{1 - \rho^2}{2} \sum_{j=1}^m \{H_0, H_j\}^2 \leq 0, \tag{4.44}$$

which is obviously satisfied for $\rho \geq 1$. Furthermore, $\rho = 1$ even solves the associated HJBe (2.10) and hence together with the assumption for the zero-state observability it is proved that

$$u_{C,j} = -\{H_j, H_0\} = -\frac{\mathrm{d}}{\mathrm{d}t}y_j, \quad j = 1, \ldots, m \tag{4.45}$$

solves the optimal non-linear H_2-control problem. ∎

4.2.4 Non-linear H_∞-design for PCH-systems

For the non-linear H_∞-design we consider a PCH-system with the Hamiltonian function

$$H = H_0 - \sum_{j=1}^{m} H_{u,j}u_{C,j} - \sum_{j=1}^{m} H_{d,j}d_j, \tag{4.46}$$

where the number of disturbance inputs d_j equals the number of control inputs $u_{C,j}$, $j = 1, \ldots, m$. Again a perfect sensor/actuator collocation is taken for granted, i.e. the output functions read as

$$y_j = H_{u,j}, \quad j = 1, \ldots, m . \tag{4.47}$$

As already discussed in detail in Section 2.3, within the scope of the non-linear H_∞-design, we are looking for an optimal solution u_C^* and d^* of an optimization problem with the objective function

$$J_\infty = \sup_{d \in L_2^m[0,\infty)} \inf_{u_C \in L_2^m[0,\infty)} \frac{1}{2} \int_0^\infty \left(\left\| \frac{\mathrm{d}}{\mathrm{d}t}y \right\|^2 + \|u_C\|^2 - \gamma \|d\|^2 \right) \mathrm{d}t \tag{4.48}$$

for the disturbance attenuation level $\gamma > 0$. In a second step, we also try to find the minimum value of γ.

Proposition 4.2. *Let us assume a PCH-system with the Hamiltonian function (4.46) and the associated natural outputs (4.47). Suppose that the control inputs act in the same way on the structure as the disturbance inputs, i.e. $H_{u,j} = H_{d,j} = H_j$, $j = 1, \ldots, m$ and that the relation $\{H_j, H_i\} = 0$ holds for all $i, j = 1, \ldots, m$. Then the control law*

$$u_{C,j} = -\sqrt{\frac{\gamma}{\gamma - 1}}\frac{\mathrm{d}}{\mathrm{d}t}y_j, \quad j = 1, \ldots, m \tag{4.49}$$

solves for $\gamma > 1$ the optimal non-linear H_∞-control problem with respect to the objective function (4.48) [118], [122], [123].

Proof. For infinite-dimensional PCH-systems recall the stability hypothesis of Subsection 4.2.2. Following Theorem 2.2 the non-linear H_∞-control problem is solved by finding a positive (semi)-definite solution V of the associated HJBIi (2.46)

$$\sup_{d\in L_2^m[0,\infty)} \inf_{u_C\in L_2^m[0,\infty)} \left(\frac{d}{dt}V + \frac{1}{2}\left(\left\|\frac{d}{dt}y\right\|^2 + \|u_C\|^2 - \gamma\|d\|^2 \right) \right) \leq 0 .$$

(4.50)

Taking into account the assumption $H_{u,j} = H_{d,j} = H_j$, $j = 1,\ldots,m$, we can rewrite (4.50) in the Poisson bracket form

$$\sup_{d\in L_2^m[0,\infty)} \inf_{u_C\in L_2^m[0,\infty)} \left(\{V,H_0\} - \sum_{j=1}^m \{V,H_j\}(u_{C,j}+d_j) + \right.$$
$$\left. \frac{1}{2}\sum_{j=1}^m \left(\{H_j,H_0\}^2 + u_{C,j}^2 - \gamma d_j^2 \right) \right) \leq 0 .$$

(4.51)

With the optimal solutions $u_{C,j}^*$ and d_j^*, $j = 1,\ldots,m$ (see also (2.44))

$$u_{C,j}^* = \{V,H_j\} \quad \text{and} \quad d_j^* = -\frac{1}{\gamma}\{V,H_j\}$$

(4.52)

and the suitable candidate for the solution of the HJBIi $V = \rho H_0$, $\rho > 0$, we get

$$\sum_{j=1}^m \frac{1}{2}\left(1 - \rho^2\left(\frac{\gamma-1}{\gamma} \right) \right) \{H_0,H_j\}^2 \leq 0 .$$

(4.53)

Clearly, for $\gamma > 1$ and $\rho \geq \sqrt{\frac{\gamma}{\gamma-1}}$ the inequality (4.53) is fulfilled. Furthermore, by setting $\rho = \sqrt{\frac{\gamma}{\gamma-1}}$, we can even solve the associated HJBIe (2.45) and hence (4.52) becomes

$$u_{C,j} = \sqrt{\frac{\gamma}{\gamma-1}}\{H_0,H_j\} = -\sqrt{\frac{\gamma}{\gamma-1}}\frac{d}{dt}y_j .$$

(4.54)

This completes the proof.   ∎

4.2.5 PD-design for PCH-systems

Again, a PCH-system with the Hamiltonian function

$$H = H_0 - \sum_{j=1}^m H_j u_{C,j}$$

(4.55)

and the associated natural outputs $y_j = H_j$, $j = 1, \ldots, m$ form the basis of our investigations. Without restriction of generality, let the origin be an equilibrium of the free system. If we succeed in finding a control law which is derived from a potential function V_u via

$$\sum_{j=1}^{m} u_{C,j} \mathrm{d} H_j = -\mathrm{d} V_u, \tag{4.56}$$

then the closed-loop can be considered as a free PCH-system with the new corresponding Hamiltonian function $\bar{H}_0 = H_0 + V_u$ [103], [119]. A well known controller, which fits this framework, is a P(proportional)-controller

$$u_{C,j} = -\sum_{i=1}^{m} P_{ji} y_i \tag{4.57}$$

with the positive (semi)-definite matrix P and the associated potential function

$$V_u = \frac{1}{2} \sum_{j=1}^{m} \sum_{i=1}^{m} P_{ji} y_i y_j . \tag{4.58}$$

In the sense of the previous subsections the control law (4.57) can be extended by a D(differential)-controller of the form

$$u_{C,j} = -\sum_{i=1}^{m} D_{ji} \frac{\mathrm{d}}{\mathrm{d}t} y_i \tag{4.59}$$

with a positive (semi)-definite matrix D. Then the change of the extended Hamiltonian function $\bar{H}_0$ due to the motion of the system is given by

$$\frac{\mathrm{d}\bar{H}_0}{\mathrm{d}t} = \{\bar{H}_0, H\} = -\sum_{j=1}^{m} \sum_{i=1}^{m} \{H_j, H\} D_{ji} \frac{\mathrm{d}}{\mathrm{d}t} y_i = -\sum_{j=1}^{m} \sum_{i=1}^{m} D_{ji} \frac{\mathrm{d}}{\mathrm{d}t} y_j \frac{\mathrm{d}}{\mathrm{d}t} y_i, \tag{4.60}$$

which is obviously less equal zero. Furthermore, if $\bar{H}_0$ is a positive definite functional in the infinite-dimensional case, then the stability hypothesis of Subsection 4.2.2 together with (4.60) implies the stability of the closed-loop system.

4.2.6 Disturbance Compensation for PCH-systems

In general, disturbance signals are considered to be signals that cannot be measured. In many cases we simply make some assumptions about the disturbance, as e.g., in the non-linear H_∞-control, where the disturbance d has to meet the requirements $d \in L_2[0, \infty)$. But if we find a way to measure directly or indirectly the disturbance, we will, of course, use this information to design the controller. Let us assume a PCH-system with the Hamiltonian function (4.46) and the corresponding Hamiltonian vector field X_H analogous to (4.34)

$$X_H = X_{H_0} - \sum_{j=1}^{m} u_{C,j} X_{H_{u,j}} - \sum_{j=1}^{m} d_j X_{H_{d,j}}, \qquad (4.61)$$

where the number of control inputs $u_{C,j}$ is equal to the number of disturbance inputs d_j, $j = 1, \ldots, m$. In some situations it is possible to design the sensor layers in such a way that the measured quantities y_j satisfy the relation

$$\sum_{j=1}^{m} \alpha_j y_j X_{H_{u,j}} = \sum_{j=1}^{m} \left(d_j X_{H_{d,j}} + \hat{X}_j \right), \quad \alpha_j \in R, \ j = 1, \ldots, m, \qquad (4.62)$$

so that $X_{\bar{H}} = X_{H_0} + \sum_{j=1}^{m} \hat{X}_j$ is still a Hamiltonian vector field with the associated Hamiltonian $\bar{H}$. Obviously, $X_{\bar{H}}$ and $\bar{H}$ are the evolutionary vector field and the Hamiltonian of the closed-loop, if the control law

$$u_{C,j} = -\alpha_j y_j, \quad j = 1, \ldots, m \qquad (4.63)$$

is used. From now on we will refer to this control strategy as the so-called disturbance compensation scheme. A very interesting application of this method in the infinite-dimensional case for a simply supported composite piezoelectric beam under the action of an axial support motion will be presented in Subsection 4.3.7.

4.3 Application: Control of Smart Piezoelectric Beam Structures

In this section, the main aim is to show the application of the previously formulated control strategies by means of an infinite-dimensional PCH-system, namely a piezoelectric composite beam structure. It is a fairly straightforward matter to extend the theory to cover plates, see, e.g., [123]. Further, we intend to make it clear that for infinite-dimensional systems the choice of the actuators and sensors can contribute quite a lot to the controller design and hence should be regarded as an integral part of the control task. In this way, it will be possible to measure those integral quantities by means of the sensors that are required for the realization of the distributed feedback laws. The actuators will provide the correctly distributed control input. In recent years, many control techniques which take into account the distributed nature of structural systems have been reported. We will not consider control concepts based on a finite approximation of the infinite-dimensional system. It should be emphasized that the distributed nature of the sensors and actuators is the fundamental idea behind all the considerations. Nevertheless, the literature contains many theoretical and/or application-driven contributions dealing with finite-dimensional approximative models which are more or less successful. To our knowledge, in [6], was the first time a Lyapunov controller for a cantilever beam, using a spatially uniform distributed actuator, was introduced. Since then other controllers based on Lyapunov's theory have been developed. Examples are the modal-filtering concept in space of [95] and different modal sensors/actuators with non-uniform spatial distributions for certain beams and plates [9], [75], [98], [139], [140].

4.3.1 Preliminaries

The field of smart structures is manifold. It ranges from aerospace applications to structural acoustics to micro-mechanical devices. In literature one can find various formulations of what smart structures are. Here we follow the definition of [4]: *Smart structures are structures or structural components on which are attached or in which are embedded sensors and actuators whose actions are coordinated through a control system imbuing the structure in proportion to their magnitudes to compensate for undesired effects or to enhance desired effects.* Apart from the piezoelectric material, which will be discussed in the following pages, other materials such as shape memory alloys (SMAs), electrorheological, electro- and magnetostrictive materials are used in smart structures. The choice of smart material for a specific application depends on many considerations, like stiffness, weight, brittleness, specific energy, temperature sensitivity, integrability, constructive flexibility etc.. It is not determined exclusively by the demands on the controller design. The direct piezoelectric effect is the phenomenon which in response to mechanical strain the piezoelectric material produces dielectric polarization. This effect is responsible for the sensor capabilities of the piezoelectric structures. Conversely, the actuating capabilities are due to the converse piezoelectric effect, which says that an electric field applied to the piezoelectric material induces a mechanical stress in the material. The most frequently used piezoelectric materials are piezoceramics (e.g., PZT's) and piezopolymers (e.g., PVDF's). Where generally piezoceramics suffer from inherent brittleness, piezopolymers have a weaker electromechanical coupling coefficient [9]. From the control point of view piezoelectric materials have the big advantage of making it easy to adjust the sensors and actuators to the special application of the structure that is to be controlled. For instance, this can be achieved by shaping the surface electrode of the piezoelectric material in the spatial domain [50], [51], [76].

It is well known that at higher electric field strengths the polarization of the piezoelectric material saturates and hence a significant hysteresis and

strain-based non-linearities appear. A tracking control strategy for a piezoceramic actuator with hysteresis can be found, e.g. in [34]. However, in context
with the piezoelectric structures which are being considered the mathematical model is restricted to the fundamental relations of linear piezoelectricity
with the constitutive equations

$$
\begin{aligned}
\sigma_{ij} &= \sum_{kl} c_{ijkl}\varepsilon_{kl} - \sum_k a_{kij}D_k \\
E_i &= -\sum_{kl} a_{ikl}\varepsilon_{kl} + \sum_k d_{ik}D_k \ ,
\end{aligned}
\tag{4.64}
$$

where σ denotes the stress, ε the strain, D the electric flux density and E
the electric field strength [105]. The integrability conditions

$$
c_{ijkl} = c_{jikl} = c_{ijlk} = c_{klij}, \ a_{kij} = a_{kji}, \ d_{ij} = d_{ji}
\tag{4.65}
$$

guarantee the existence of an energy function w_p

$$
w_p = \int_0^t \int_V \left(\sum_{ij} \sigma_{ij}\frac{\partial}{\partial t}\varepsilon_{ij} + \sum_i E_i\frac{\partial}{\partial t}D_i \right) w_v dt
\tag{4.66}
$$

or

$$
w_p = \int_V \frac{1}{2}\left(\sum_{ijkl} c_{ijkl}\varepsilon_{kl}\varepsilon_{ij} - 2\sum_{ikl} a_{ikl}\varepsilon_{kl}D_i + \sum_{ik} d_{ik}D_kD_i \right) w_v
\tag{4.67}
$$

with V as the volume of the structure and $w_v = dz_1 dz_2 dz_3$ as the related
volume element. Here, we assume that no energy is stored in the structure at
time $t = 0$, i.e. initially $w_p = 0$. Many piezoelectric materials are relatively
insensitive to temperature variations [9]. For this reason we shall also ignore
the coupling with the thermal field in the constitutive equations (4.64). A
mathematical model of piezoelectric sensors and actuators, which takes all
the effects of the interaction of mechanical, electrical and thermal fields into
account, is presented in [49]. Inside the piezoelectric lamina, the free volume
charge density is zero and hence Maxwell's equation for D reads as

$$
\sum_i \frac{\partial}{\partial z_i}D_i = 0 \ .
\tag{4.68}
$$

In an electrostatic field the electric potential P_{el} completely describes the
electric field strength E by

$$
E_i = -\frac{\partial}{\partial z_i}P_{el}
\tag{4.69}
$$

and since the metallic electrodes define equipotential surfaces, the electric
field strength E is perpendicular to the electrodes.

4.3.2 Beam Structure under Consideration

Generally, a piezoelectric beam consists of a large number of thin layers of laminae with and without piezoelectric properties. The different piezoelectric layers are supposed to be perfectly bonded to the substrate and they can be either used as actuators or sensors. Henceforth, all our considerations are based on a simply supported piezoelectric composite beam shown in Fig. 4.1. The flexural vibrations due to different disturbances, namely lateral loadings and axial support motions, are studied in the (z_1, z_3)-plane, where z_1 is the axial and z_3 the lateral coordinate. The longitudinal displacement will be denoted by u_1 and the normal displacement by u_3.

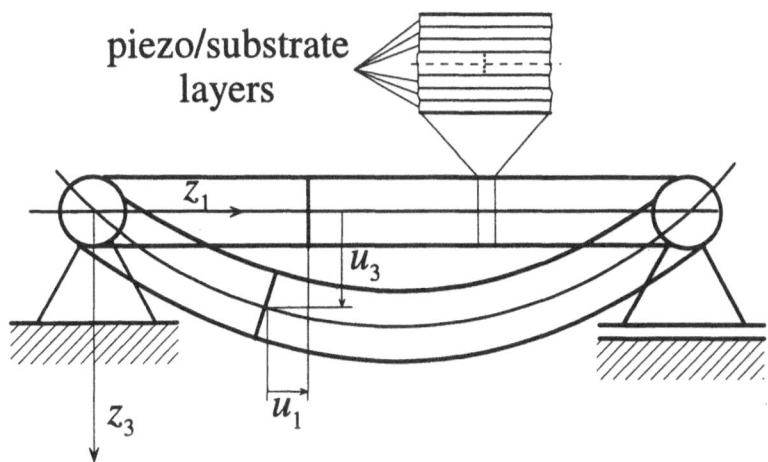

Fig. 4.1. Simply supported composite piezoelectric beam.

For the derivation of the mathematical model let us assume that the stresses $\sigma_{ij} = 0$ for $i + j > 2$ and the electric flux density $D_i = 0$ for $i = 1, 2$. Since (4.68) must hold, it follows that the component D_3 of the electric flux density is a function of z_1 and z_2. In the literature one can find other approaches where assumptions on the components of the electric field strength E are made. Essentially, this approach leads to the same structure of the mathematical model and hence does not change anything for the controller design. Inserting these simplifying assumptions into (4.64), we obtain the constitutive equations of the beam in the form

$$\sigma_{11} = c^* \varepsilon_{11} - a^* D_3 \tag{4.70}$$

and

$$E_3 = -a^* \varepsilon_{11} + d^* D_3 \tag{4.71}$$

with the effective material parameters a^*, c^* and d^*. The longitudinal strain ε_{11} is related to the beam's curvature by means of the Bernoulli-Euler assumption

$$\varepsilon_{11} = \bar{\varepsilon}_{11} - z_3 \frac{\partial^2}{\partial z_1^2} u_3 \, , \tag{4.72}$$

where $\bar{\varepsilon}_{11}$ is the strain in the axis $z_3 = 0$ (e.g., [147]). In the sense of v. Karman a non-linear formulation is used for $\bar{\varepsilon}_{11}$, i.e.

$$\bar{\varepsilon}_{11} = \frac{\partial}{\partial z_1} u_1 + \frac{1}{2} \left(\frac{\partial}{\partial z_1} u_3 \right)^2 \, . \tag{4.73}$$

Thus, the potential energy (4.67) for the beam being considered takes the form

$$w_p = \frac{1}{2} \int_V \left(c^* \varepsilon_{11}^2 - 2a^* \varepsilon_{11} D_3 + d^* D_3^2 \right) w_v =$$

$$\frac{1}{2} \int_V \left(c^* \left(\bar{\varepsilon}_{11} - z_3 \frac{\partial^2}{\partial z_1^2} u_3 \right)^2 - 2a^* \left(\bar{\varepsilon}_{11} - z_3 \frac{\partial^2}{\partial z_1^2} u_3 \right) D_3 + d^* D_3^2 \right) w_v \, . \tag{4.74}$$

By neglecting longitudinal and rotational inertia, the kinetic energy w_k is given by

$$w_k = \int_0^L \frac{\mu}{2} \left(\frac{\partial}{\partial t} u_3 \right)^2 dz_1 \tag{4.75}$$

with $\mu = \int_A \rho w_a$, the mass density ρ, the cross section of the beam A lying in the plane $z_1 = \text{const.}$ and the related area element $w_a = dz_2 dz_3$.

4.3.3 Actuator and Sensor Design

Following the discussion in Subsection 4.3.1, the piezoelectric structures allow a spatial shaping of the piezoelectric layers, which can be used as an additional degree of freedom for the controller design. Before starting with the actual actuator and sensor design, we will summarize the contributions, which a suitable choice for the actuators and sensors can make to the controller design, in the following remark.

Remark 4.5. For this purpose let us assume that the piezoelectric structure is an infinite-dimensional PCH-system with the Hamiltonian functional (see (4.46))

$$H = H_0 - \sum_{j=1}^m H_{u,j} u_{C,j} - \sum_{j=1}^l H_{d,j} d_j \, . \tag{4.76}$$

- At first, let the disturbance inputs d_j, $j = 1, \ldots, l$ be zero. If we succeed in designing the piezoelectric sensor and actuator layers in such a way that the associated natural outputs $y_j = H_{u,j}$, $j = 1, \ldots, m$, are measured, then the non-linear H_2-design of Subsection 4.2.3 and the PD-controller design of Subsection 4.2.5 can be solved.
- In the case of nonvanishing disturbances, if it is possible to design $l = m$ piezoelectric actuator layers such that the relation $H_{u,j} = H_{d,j} = H_j$, $j = 1, \ldots, m$, holds and the corresponding sensor layers measure the natural outputs H_j, $j = 1, \ldots, m$, then we can find a solution for the non-linear H_∞-design of Subsection 4.2.4. The condition $H_{u,j} = H_{d,j}$ says that the plant input u_j (in our case the voltage applied to the j'th actuator layer) acts in the same way on the structure as the disturbance d_j, with the spatial distribution $H_{d,j}$. Therefore, if d_j is known, its influence on the system can be exactly eliminated [50], [51].
- In some cases the design of the piezoelectric actuator and sensor layers enables a disturbance compensation in the sense of Subsection 4.2.6.

Without going into the details of the special realization, we will subsequently elaborate the fundamental possibilities of designing a spatially distributed piezoelectric actuator and sensor layer. The actuator design is based on the constitutive equation (4.71) where we neglect the influence of ε_{11} on E_3. This simplification is admissible, since the self-generated voltage due to the direct piezoelectric effect is insignificant compared with the applied electric field (see e.g., [137]). The integration along the electric field strength E_3 from one electrode of the j'th piezoelectric layer to the other leads to the result

$$\int_{h_{1,j}(z_1)}^{h_{2,j}(z_1)} E_3 dz_3 = d^* \left(h_{2,j}(z_1) - h_{1,j}(z_1) \right) D_3 = d^* T_j(z_1) D_3 = U_j \qquad (4.77)$$

or

$$D_3 = \frac{U_j}{d^* T_j(z_1)} \qquad (4.78)$$

with the height of the piezoelectric lamina $T_j(z_1)$ and the applied voltage U_j. Now we assume that the beam is built up symmetrically with respect to the mid-plane $z_3 = 0$. But the voltage U_j and hence the poling field applied to the two piezoelectric layers of the j'th layer couple can be chosen to be symmetric or antisymmetric with respect to $z_3 = 0$. Thus, let us consider a piezoelectric beam with $2m$ layers, where m_a layer couples are supplied antisymmetrically by a voltage U_j^a, $j = 1, \ldots, m_a$ and m_s layer couples are supplied symmetrically by a voltage U_j^s, $j = 1, \ldots, m_s$. From now on, the symbol s stands for symmetric and a for antisymmetric. Apart from supplying the two piezoelectric layers of the j'th layer couple with an antisymmetric voltage U_j^a, one can obtain the same results by varying the poling direction of the piezoelectric lamina accordingly.

Under the assumption that the voltage sources for the supply voltages U_j^s and U_j^a are ideal we can neglect the term $d^* D_3^2$ in (4.74). Consequently, taking into account (4.78) and the symmetry of the effective material parameters a^*, c^* and d^* with respect to $z_3 = 0$, we obtain the potential energy (4.74) of the beam in the form

$$
w_p = \int_0^L \frac{1}{2} \left(\Lambda_1 \left(\frac{\partial}{\partial z_1} u_1 + \frac{1}{2} \left(\frac{\partial}{\partial z_1} u_3 \right)^2 \right)^2 + \Lambda_2 \left(\frac{\partial^2}{\partial z_1^2} u_3 \right)^2 \right) dz_1 -
$$

$$
\int_0^L \sum_{j=1}^{m_s} \Lambda_j^s (z_1) \left(\frac{\partial}{\partial z_1} u_1 + \frac{1}{2} \left(\frac{\partial}{\partial z_1} u_3 \right)^2 \right) U_j^s dz_1 +
$$

$$
\int_0^L \sum_{j=1}^{m_a} \Lambda_j^a(z_1) \left(\frac{\partial^2}{\partial z_1^2} u_3 \right) U_j^a dz_1
$$

(4.79)

with

$$
\Lambda_1 = \sum_{j=1}^{2m} \int_{A_j} c_j^* \omega_a \quad , \quad \Lambda_2 = \sum_{j=1}^{2m} \int_{A_j} c_j^* z_3^2 \omega_a
$$

(4.80)

and

$$
\Lambda_j^s(z_1) = \int_{h_{1,j}^s(z_1)}^{h_{2,j}^s(z_1)} \int_{b_{1,j}^s(z_1)}^{b_{2,j}^s(z_1)} \frac{2 a_j^*}{d_j^* T_j^s(z_1)} \omega_a = \frac{2 a_j^*}{d_j^*} B_j^s (z_1)
$$

$$
\Lambda_j^a(z_1) = \int_{h_{1,j}^a(z_1)}^{h_{2,j}^a(z_1)} \int_{b_{1,j}^a(z_1)}^{b_{2,j}^a(z_1)} \frac{2 a_j^* z_3}{d_j^* T_j^a(z_1)} \omega_a = \frac{a_j^* \left(h_{2,j}^a(z_1) + h_{1,j}^a(z_1) \right) B_j^a (z_1)}{d_j^*} .
$$

(4.81)

A_j denotes the cross section of the j'th layer and $\omega_a = dz_2 dz_3$ is the related area element. Here, $\Lambda_j^s(z_1)$ and $\Lambda_j^a(z_1)$ serve as shaping functions which can be adjusted according to the requirements of the controller design [68]. At the edges of the electrodes $b_{1,j}^s(z_1)$, $b_{2,j}^s(z_1)$, $b_{1,j}^a(z_1)$ and $b_{2,j}^a(z_1)$ the electric field is assumed to be homogenous and no edge effects are taken into account. Hence, inside the structural lamina (from $h_{1,j}^s(z_1)$ to $h_{2,j}^s(z_1)$ or from $h_{1,j}^a(z_1)$ to $h_{2,j}^a(z_1)$) and outside the area of the piezoelectric lamina not covered by the electrodes, the voltages U_j^s and U_j^a are zero. Fig. 4.2 sketches a possibility of creating a specified spatial distribution of $\Lambda_j^s(z_1)$ and $\Lambda_j^a(z_1)$ by means of shaping the corresponding electrodes. At this point it should be explained that the poling direction in the piezoelectric layer can only be up or down, due to polarity. The voltage supplied, U_j, is either positive or negative. Fig. 4.2 shows all different possible combinations for creating a symmetrically or antisymmetrically supplied piezoelectric layer couple. Fig. 4.3 shows a second possibility for $\Lambda_j^a(z_1)$, where the thickness of the piezoelectric lamina varies over the length of the layer. Of course, a combination of these methods is also possible. It should be emphasized that Figs. 4.2 and 4.3 depict only the

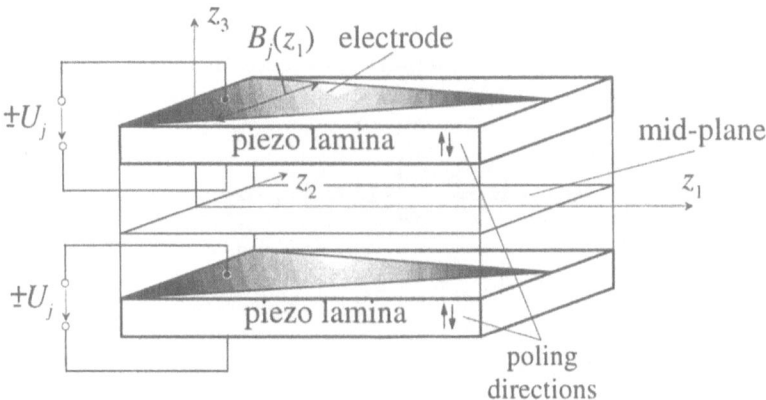

Fig. 4.2. Principle of surface shaping of the electrode for an actuator layer couple.

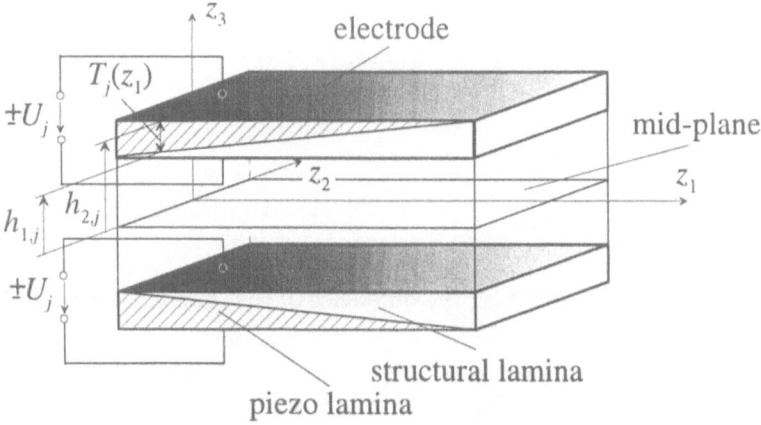

Fig. 4.3. Principle of shaping the piezoelectric lamina for an actuator layer couple.

ideas relating to the design of a specified shaping function in principle [68]. In a practical application one will use more sophisticated surface patterns of the electrodes for achieving the shaping functions (see e.g., [75] and the references cited there).

Analogous to the actuator design, the constitutive equation (4.71) also serves as a basis for the derivation of the sensor equations. But here we assume that the electrodes of a piezoelectric sensor layer are short circuited, i.e.

$$-a^* \varepsilon_{11} + d^* D_3 = 0 , \tag{4.82}$$

and in addition they are assumed to have an identical surface pattern. By integration over the effective metallic surface of the electrodes of the j'th sensor layer we get the electric charge Q_j in the form [75]

$$Q_j = \int_0^L \int_{b_{1,j}(z_1)}^{b_{2,j}(z_1)} D_3 dz_2 dz_1 = \int_0^L \frac{a_j^*}{d_j^*} B_j(z_1) \varepsilon_{11} dz_1 \tag{4.83}$$

or

$$Q_j = \int_0^L \frac{a_j^*}{d_j^*} B_j(z_1) \left(\frac{\partial}{\partial z_1} u_1 + \frac{1}{2} \left(\frac{\partial}{\partial z_1} u_3 \right)^2 - z_3 \frac{\partial^2}{\partial z_1^2} u_3 \right) dz_1 \tag{4.84}$$

with $\bar{z}_3$ as the distance from the mid-plane to the middle of the j'th sensor layer (see Fig. 4.4). Since the layers of the piezoelectric beam are arranged symmetrically with respect to the mid-plane, we again have two possibilities for measuring the charge. On the one hand we can take the sum of the charge of the two corresponding layers of a sensor layer couple and we get

$$Q_j^s = \int_0^L \Gamma_j^s(z_1) \left(\frac{\partial}{\partial z_1} u_1 + \frac{1}{2} \left(\frac{\partial}{\partial z_1} u_3 \right)^2 \right) dz_1 \tag{4.85}$$

with the shaping function

$$\Gamma_j^s(z_1) = \frac{2a_j^*}{d_j^*} B_j(z_1) . \tag{4.86}$$

On the other hand, by taking the difference of the charge, we directly obtain the result

$$Q_j^a = \int_0^L \Gamma_j^a(z_1) \frac{\partial^2}{\partial z_1^2} u_3 dz_1 \tag{4.87}$$

with the shaping function

$$\Gamma_j^a (z_1) = \frac{2\bar{z}_3 a_j^*}{d_j^*} B_j (z_1) \ . \tag{4.88}$$

By means of these shaping functions it is possible to measure specified spatially distributed quantities [68]. The principle of shaping the sensor layers is shown in Fig. 4.4. Here also the right choice of the polarization profile within each layer of one sensor layer couple offers an additional possibility to create the shaping functions (4.86) and (4.88).

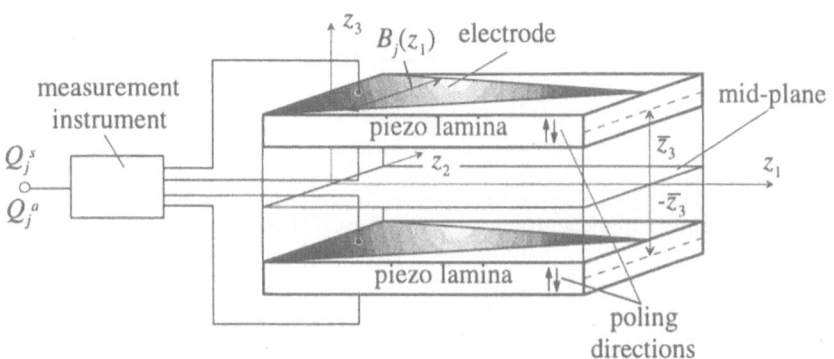

Fig. 4.4. Principle of surface shaping of the electrode for a sensor layer couple.

4.3.4 Mathematical Model for the Beam with Lateral Loadings

Fig. 4.5 depicts the piezoelectric beam of Fig. 4.1 under the action of two lateral loadings, one of which is assumed to be constant spatially $\chi_1 (z_1, t) = d_1 (t)$. The other is linear $\chi_2 (z_1, t) = d_2 (t) z_1 / L$ with L as the total length of the beam [68]. By running through the Hamilton formalism with the kinetic energy of (4.75) and the potential energy of (4.79), we obtain the equations of motion in the form

$$\mu \frac{\partial^2}{\partial t^2} u_3 + \Lambda_2 \frac{\partial^4}{\partial z_1^4} u_3 - \Lambda_1 \frac{\partial}{\partial z_1} \left\{ \left(\frac{\partial}{\partial z_1} u_1 + \frac{1}{2} \left(\frac{\partial}{\partial z_1} u_3 \right)^2 \right) \frac{\partial}{\partial z_1} u_3 \right\} +$$

$$\sum_{j=1}^{m_s} \frac{\partial}{\partial z_1} \left(\Lambda_j^s (z_1) \frac{\partial}{\partial z_1} u_3 \right) U_j^s + \sum_{j=1}^{m_a} \frac{\partial^2}{\partial z_1^2} \Lambda_j^a (z_1) U_j^a - \chi = 0$$

$$\tag{4.89}$$

and

$$\Lambda_1 \frac{\partial}{\partial z_1} \left(\frac{\partial}{\partial z_1} u_1 + \frac{1}{2} \left(\frac{\partial}{\partial z_1} u_3 \right)^2 \right) - \sum_{j=1}^{m_s} \frac{\partial}{\partial z_1} \Lambda_j^s (z_1) U_j^s = 0. \tag{4.90}$$

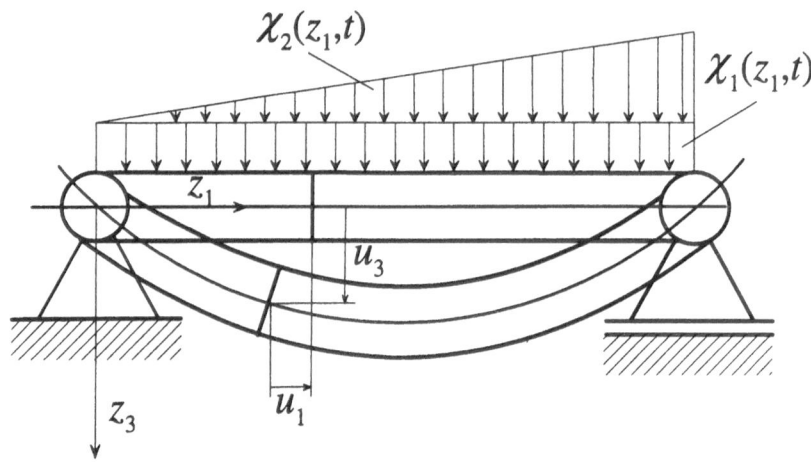

Fig. 4.5. Simply supported composite piezoelectric beam under the action of lateral loadings.

Here, $\chi = d_1(t) + d_2(t) z_1/L$ denotes the lateral loading as shown in Fig. 4.5. The flexural boundary conditions for the beam, which is simply supported at $z_1 = 0$ and $z_1 = L$, are

$$u_3 = 0 \quad \text{and} \quad \Lambda_2 \frac{\partial^2}{\partial z_1^2} u_3 + \sum_{j=1}^{m_a} \Lambda_j^a(z_1) U_j^a = 0 \tag{4.91}$$

and the longitudinal boundary conditions are given by

$$u_1(0,t) = u_1(L,t) = 0 . \tag{4.92}$$

Remark 4.6. At this point it is worth mentioning that the theory being presented also comprises beams with other boundary conditions, like e.g. a cantilever beam. All subsequent considerations remain valid and can be applied without additional effort (see, e.g., [40], [123]).

The evaluation of the integral (see (4.90))

$$\int_0^L \left(\Lambda_1 \left(\frac{\partial}{\partial z_1} u_1 + \frac{1}{2} \left(\frac{\partial}{\partial z_1} u_3 \right)^2 \right) - \sum_{j=1}^{m_s} \Lambda_j^s(z_1) U_j^s \right) dz_1 =$$

$$\left(\Lambda_1 \left(\frac{\partial}{\partial z_1} u_1 + \frac{1}{2} \left(\frac{\partial}{\partial z_1} u_3 \right)^2 \right) - \sum_{j=1}^{m_s} \Lambda_j^s(z_1) U_j^s \right) L =$$

$$= \int_0^L \left(\frac{\Lambda_1}{2} \left(\frac{\partial}{\partial z_1} u_3 \right)^2 - \sum_{j=1}^{m_s} \Lambda_j^s (z_1) U_j^s \right) dz_1 + \Lambda_1 \left(u_1 \left(L, t \right) - u_1 \left(0, t \right) \right)$$

(4.93)

together with (4.92) leads to a simplification of (4.89) in the following form

$$\mu \frac{\partial^2}{\partial t^2} u_3 + \Lambda_2 \frac{\partial^4}{\partial z_1^4} u_3 + \sum_{j=1}^{m_a} \frac{\partial^2}{\partial z_1^2} \Lambda_j^a(z_1) U_j^a - \left(d_1 \left(t \right) + d_2 \left(t \right) \frac{z_1}{L} \right) -$$

$$\left\{ \frac{1}{L} \int_0^L \left(\frac{\Lambda_1}{2} \left(\frac{\partial}{\partial z_1} u_3 \right)^2 - \sum_{j=1}^{m_s} \Lambda_j^s (z_1) U_j^s \right) dz_1 \right\} \frac{\partial^2}{\partial z_1^2} u_3 = 0 \,.$$

(4.94)

For the sake of simplicity of the notation the problem-oriented scaling with respect to the total height T of the beam

$$
\begin{aligned}
\tilde{z} &= \frac{z_1}{L}, & \tilde{u}_3 &= \frac{u_3}{T}, & \tilde{t} &= \frac{t}{L^2} \sqrt{\frac{\Lambda_2}{\mu}}, \\
\tilde{k} &= \frac{\Lambda_1 T^2}{\Lambda_2}, & \tilde{\Lambda}_j^{a,s} (\tilde{z}) &= \Lambda_j^{a,s} (z_1), & \tilde{u}_1 &= u_1 \frac{L}{T^2}, \\
\tilde{d}_1 &= d_1 \frac{L^4}{\Lambda_2 T}, & \tilde{d}_2 &= d_2 \frac{L^4}{\Lambda_2 T}, & \tilde{u}_{C,j}^s &= U_j^s \frac{L^2}{\Lambda_1 T^2}, \\
\tilde{u}_{C,j}^a &= U_j^a \frac{L^2}{\Lambda_2 T}
\end{aligned}
$$

(4.95)

is introduced, where a tilde refers to a non-dimensional quantity. Henceforth, we will ignore the tilde-symbol. Thus, with the abbreviation $p = \frac{\partial}{\partial t} u_3$ the equations of motion take the form

$$\frac{\partial}{\partial t} p + \frac{\partial^4}{\partial z^4} u_3 + \sum_{j=1}^{m_a} \frac{\partial^2}{\partial z^2} \Lambda_j^a(z) u_{C,j}^a - \left(d_1 \left(t \right) + d_2 \left(t \right) z \right) -$$

$$k \left\{ \int_0^1 \left(\frac{1}{2} \left(\frac{\partial}{\partial z} u_3 \right)^2 - \sum_{j=1}^{m_s} \Lambda_j^s (z) u_{C,j}^s \right) dz \right\} \frac{\partial^2}{\partial z^2} u_3 = 0$$

(4.96)

with the boundary conditions

$$u_3 = 0 \quad \text{and} \quad \frac{\partial^2}{\partial z^2} u_3 + \sum_{j=1}^{m_a} \Lambda_j^a(z) u_{C,j}^a = 0 \quad \text{for} \quad z \in \{0, 1\} \,.$$

(4.97)

In fact, (4.96) is an infinite-dimensional PCH-system with the Hamiltonian functional

$$H = H_0 - \sum_{j=1}^{m_a} H_{u,j}^a u_{C,j}^a - \sum_{j=1}^{m_s} H_{u,j}^s u_{C,j}^s - \sum_{j=1}^{2} H_{d,j} d_j \,,$$

(4.98)

where the Hamiltonian functional of the free system H_0 is given by

$$H_0 = \frac{1}{2} \int_0^1 \left(p^2 + \left(\frac{\partial^2}{\partial z^2} u_3 \right)^2 \right) dz + \frac{k}{8} \left(\int_0^1 \left(\frac{\partial}{\partial z} u_3 \right)^2 dz \right)^2 \tag{4.99}$$

and the interaction Hamiltonian functionals read as

$$H_{u,j}^a = - \int_0^1 \Lambda_j^a(z) \frac{\partial^2}{\partial z^2} u_3 dz \quad , j = 1, \ldots, m_a$$

$$H_{u,j}^s = \frac{k}{2} \int_0^1 \left(\int_0^1 \Lambda_j^s(z) dz \right) \left(\frac{\partial}{\partial z} u_3 \right)^2 dz \quad , j = 1, \ldots, m_s \tag{4.100}$$

$$H_{d,j} = \int_0^1 u_3 z^{(j-1)} dz \quad , j = 1, 2 .$$

Consequently, we can also formulate the measured charges of (4.85) and (4.87) in the scaled version with

$$\tilde{\Gamma}_j^s (z) = \Gamma_j^s (z_1), \quad \tilde{Q}_j^s = Q_j^s \frac{L}{T^2}$$

$$\tilde{\Gamma}_j^a (z) = \Gamma_j^a (z_1), \quad \tilde{Q}_j^a = Q_j^a \frac{L}{T} . \tag{4.101}$$

By neglecting the tilde again, we obtain

$$Q_j^s = \int_0^1 \Gamma_j^s (z) \left(\frac{\partial}{\partial z} u_1 + \frac{1}{2} \left(\frac{\partial}{\partial z} u_3 \right)^2 \right) dz \tag{4.102}$$

and

$$Q_j^a = \int_0^1 \Gamma_j^a (z) \frac{\partial^2}{\partial z^2} u_3 dz . \tag{4.103}$$

4.3.5 Controller Design for the Beam with Lateral Loadings

In Remark 4.5 we have pointed out that the design of the sensors and ac-tuators for infinite-dimensional PCH-systems can be regarded as a part of the controller synthesis. We have further shown that the application of the non-linear H_∞-controller design of Proposition 4.2 is only possible if for the underlying PCH-system the control inputs act in the same way on the struc-ture as the disturbance inputs, i.e. $H_{u,j} = H_{d,j} = H_j, j = 1, \ldots, m$, and the sensors measure the natural outputs, i.e. $y_j = H_j$, $j = 1,\dots,m$. In order to meet these requirements for the mathematical model (4.96) - (4.100), we choose two antisymmetrically supplied piezoelectric actuator layer couples ($m_a = 2$, $m_s = 0$) with the shaping functions

$$\Lambda_1^a(z) = (1 - z)\frac{z}{2} \quad \text{and} \quad \Lambda_2^a(z) = (1 - z^2)\frac{z}{6}.$$ (4.104)

Inserting (4.104) into the interaction Hamiltonian functionals $H_{u,j}^a$, $j = 1,2$ of (4.100), we can immediately see by simple integration by parts

$$H_{u,j}^a = -\int_0^1 \Lambda_j^a(z)\frac{\partial^2}{\partial z^2}u_3 dz = \underbrace{-\Lambda_j^a(z)\frac{\partial}{\partial z}u_3\Big|_0^1}_{=0} + \underbrace{\frac{\partial}{\partial z}\Lambda_j^a(z)u_3\Big|_0^1}_{=0}$$

$$-\int_0^1 \frac{\partial^2}{\partial z^2}\Lambda_j^a(z)u_3 dz$$ (4.105)

that the condition $H_{u,j} = H_{d,j} = H_j$ for $j = 1,2$ is satisfied. Thus, with the shaping functions (4.104) the equations of motion (4.96) are simplified to

$$\frac{\partial}{\partial t}p + \frac{\partial^4}{\partial z^4}u_3 - k\left(\int_0^1 \frac{1}{2}\left(\frac{\partial}{\partial z}u_3\right)^2 dz\right)\frac{\partial^2}{\partial z^2}u_3 - (d_1(t) + u_{C,1}^a) -$$

$$(d_2(t) + u_{C,2}^a)z = 0$$ (4.106)

with the boundary conditions

$$u_3 = 0 \quad \text{and} \quad \frac{\partial^2}{\partial z^2}u_3 = 0 \quad \text{for} \quad z \in \{0,1\}.$$ (4.107)

Furthermore, we have to design two piezoelectric sensor layer couples to measure the required natural outputs $H_{u,j}$, $j = 1,2$. Comparing (4.103) with (4.105), we can directly deduce that by means of the shaping functions

$$\Gamma_j^a(z) = -\Lambda_j^a(z) \quad , \quad j = 1,2$$ (4.108)

the natural outputs $Q_j^a = H_{u,j}^a$, $j = 1,2$ are measured.

Simulation Model. For the purpose of simulation the deflection u_3 is approximated by the finite series

$$u_3(z,t) = \sum_{i=1}^l X_i(t)\sin(i\pi z) \quad , \quad 0 < l < \infty$$ (4.109)

and this approximation is inserted in the equations of motion (4.106). The resulting error is interpreted as a transverse loading and following the principle of Galerkin, the weighted erroneous system has to form an equilibrium system (e.g., [147]). This procedure leads to a set of non-linear ordinary differential equations

$$\frac{dX_i}{dt} = V_i$$

$$\frac{dV_i}{dt} = -X_i (i\pi)^2 \left((i\pi)^2 + \frac{k}{4} \sum_{j=1}^{l} (j\pi)^2 X_j^2 \right) + \qquad (4.110)$$

$$\frac{2}{i\pi} (1 - \cos(i\pi)) (u_{C,1}^a + d_1) - \frac{2}{i\pi} \cos(i\pi) (u_{C,2}^a + d_2)$$

for $i = 1, \ldots, l$. The finite model (4.110) is a member of a special class of finite-dimensional non-linear systems, namely the so-called AI(affine-input)-systems [52], [103], [144]. A non-linear controller design based on the theory of exact input-to-output linearization for this finite model can be found in [117].

Non-linear H_∞-design for the Beam with Lateral Loadings. Since by the actuator and sensor design of (4.104) and (4.108) the conditions for the non-linear H_∞-controller design are fulfilled, we are able to apply Proposition 4.2 to the PCH-system (4.106). Hence the optimal control law reads as

$$u_{C,j}^a = -\alpha \frac{d}{dt} Q_j^a \quad \text{with} \quad \alpha = \sqrt{\frac{\gamma}{\gamma - 1}}, j = 1, 2 \qquad (4.111)$$

for a given $\gamma > 1$. Fig. 4.6 shows the simulation results for $l = 6$, the scaled material parameter $k = 12$ and the lateral loadings $d_1(t) = 0$ and $d_2(t)$ is the sawtooth in Fig. 4.6 a.). The deflection in the middle of the beam $u_3(0.5, t)$ for the uncontrolled case $\alpha = 0$ and for the controlled case with the controller parameters $\alpha = 10$ and $\alpha = 100$ is illustrated in Fig. 4.6 b.). The corresponding voltages $u_{C,1}^a$ and $u_{C,2}^a$ are demonstrated in Figs. 4.6 c.) and d.), respectively [68]. One can see that, depending on the maximum voltage supply allowed, a better suppression of the vibrations can be achieved. Fig. 4.7 depicts the simulation results of the piezoelectric beam under the action of step changes of the lateral loadings in the form

$$d_1(t) = 5\sigma(t) \quad \text{and} \quad d_2(t) = 10\sigma(t - 3), \qquad (4.112)$$

whereby $\sigma(t)$ denotes the unit step. Since the controller (4.111) does not have an integral action, the stationary deflection caused by the step changes of the lateral loadings is not eliminated. This is why we shall apply a PD-control law to the piezoelectric beam in the next subsection.

Remark 4.7. In the configuration of two collocated piezoelectric actuator/sensor layer couples with the shaping functions (4.104) and (4.108), the vibrations caused by any lateral loading, independently of its spatial distribution,

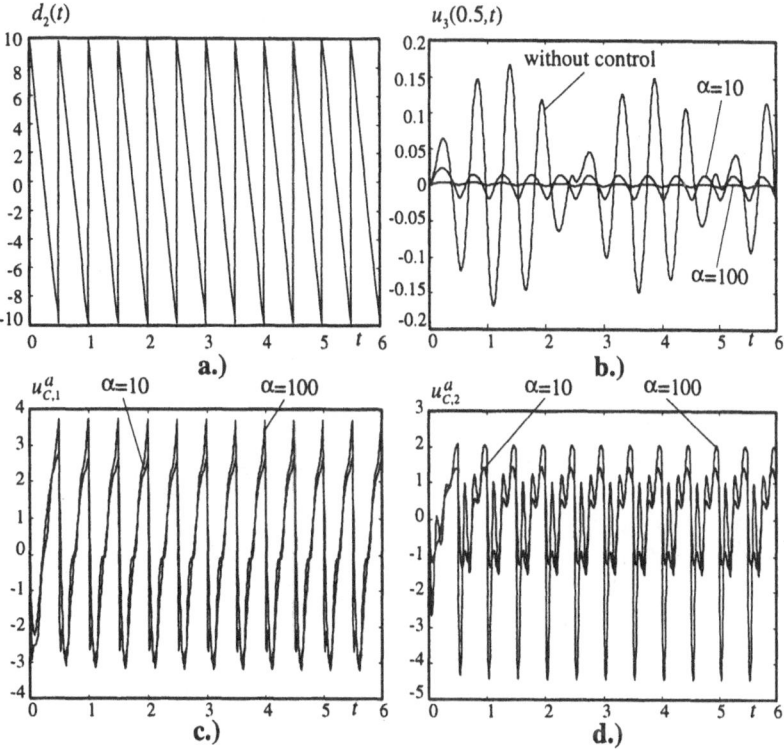

Fig. 4.6. Simulation results for the non-linear H_∞-design for the beam with time-harmonic lateral loadings.

can be suppressed. The reason is that one actuator layer couple, which acts as a spatially constant load on the structure, cancels all even vibration modes. The other actuator layer couple, which imitates a spatially linear load, rejects all odd modes. This fact can be easily observed in the finite approximation (4.110).

PD-design for the Beam with Lateral Loadings. The actuator and sensor design of the previous subsection with its inherent sensor/actuator collocation also allows us to perform a PD-controller design due to Subsection 4.2.5. Here, we restrict the general PD-feedback of (4.57) and (4.59) to

$$
\begin{bmatrix} u_{C,1}^a \\ u_{C,2}^a \end{bmatrix} = - \begin{bmatrix} P_{11} & 0 \\ 0 & P_{11} \end{bmatrix} \begin{bmatrix} Q_1^a \\ Q_2^a \end{bmatrix} - \begin{bmatrix} D_{11} & 0 \\ 0 & D_{11} \end{bmatrix} \begin{bmatrix} \dfrac{\mathrm{d}}{\mathrm{d}t} Q_1^a \\ \dfrac{\mathrm{d}}{\mathrm{d}t} Q_2^a \end{bmatrix}
\tag{4.113}
$$

with the positive constants P_{11} and D_{11} and the scaled measured charges $Q_j^a = H_{u,j}$, $j = 1, 2$. For the investigation with a sawtooth excitation (Fig.

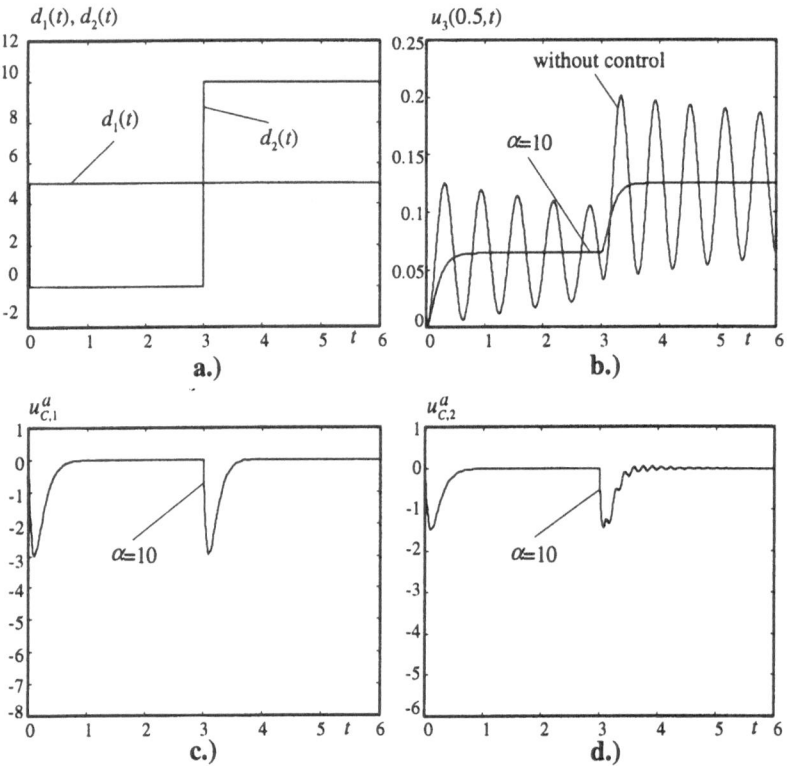

Fig. 4.7. Simulation results for the non-linear H_∞-design for the beam with step changes of the lateral loading.

4.6 a.)) we get simulation results similar to the ones of Figs. 4.6 b.)-d.). In contrast to this the PD-feedback (4.113) shows different results from the non-linear H_∞-controller for the step changes of the lateral loadings (4.112). Now, by means of the P-controller part, it is also possible to decrease the stationary error of the deflection $u_3\,(0.5, t)$. Fig. 4.8 depicts the simulation results for two different parameter sets, namely $P_{11} = 100$ and $D_{11} = 20$ in the first and $P_{11} = 1000$ with $D_{11} = 50$ in the second case. These simulations should further demonstrate that by adjusting the controller parameters P_{11} and D_{11} a certain performance of the closed-loop can be obtained, provided that the required voltages are within the possible voltage range.

Remark 4.8. However, with an integral part in the controller, the stationary error can be made zero. But an integrator in the controller not only does not fit the framework of Hamiltonian systems, it also partially destroys the Hamiltonian structure. Hence, to the best knowledge of the author, further research work is necessary to overcome these problems in the infinite-dimensional case.

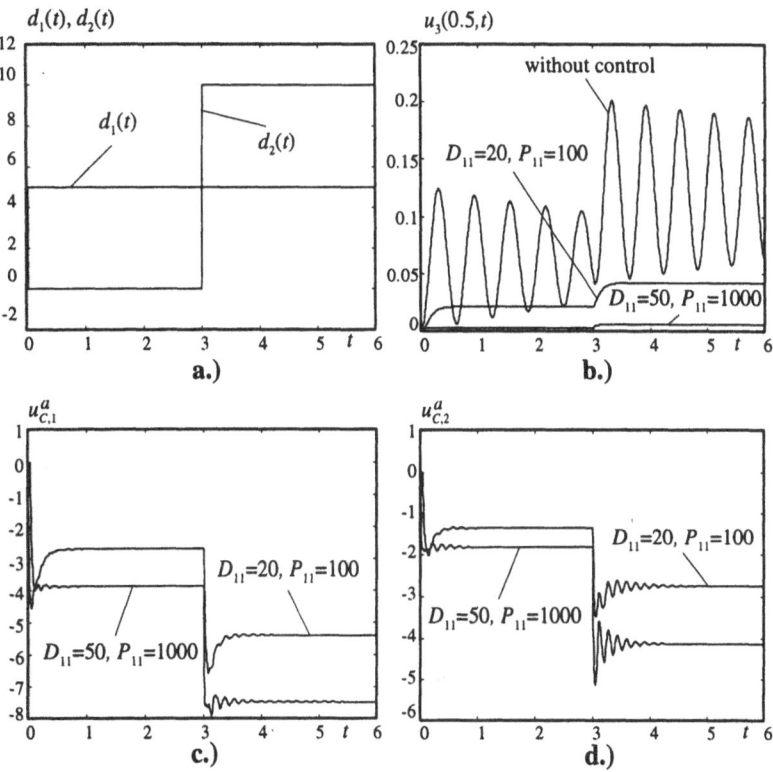

Fig. 4.8. Simulation results for the PD-design for the beam with step changes of the lateral loading.

4.3.6 Mathematical Model for the Beam with an Axial Support Motion

Now, we will discuss the simply supported composite piezoelectric beam of Fig. 4.1 under the action of an axial support motion $d_3(t)$ (see Fig. 4.9).

Except for the longitudinal boundary conditions (4.92), which change to

$$u_1(0,t) = 0 \quad \text{and} \quad u_1(L,t) = d_3(t) , \tag{4.114}$$

the mathematical model of Subsection 4.3.4 remains the same. Since the lateral loadings are zero, i.e. $d_1 = d_2 = 0$, the adapted equations of motion (4.96) and (4.97) read as

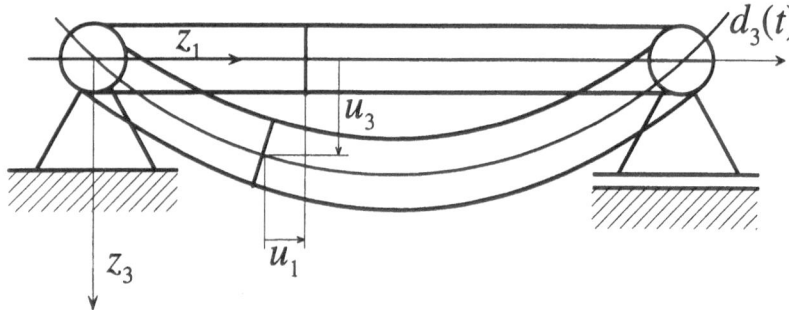

Fig. 4.9. Simply supported composite piezoelectric beam under the action of an axial support motion.

$$\frac{\partial}{\partial t}p + \frac{\partial^4}{\partial z^4}u_3 + \sum_{i=1}^{m_a}\frac{\partial^2}{\partial z^2}\Lambda_i^a(z)u_{C,i}^a - ku_1\,(1,t)\,\frac{\partial^2}{\partial z^2}u_3 -$$

$$k\int_0^1\left(\frac{1}{2}\left(\frac{\partial}{\partial z}u_3\right)^2 - \sum_{i=1}^{m_s}\Lambda_i^s(z)\,u_{C,i}^s\right)dz\frac{\partial^2}{\partial z^2}u_3 = 0$$

(4.115)

with the boundary conditions

$$u_3 = 0 \quad \text{and} \quad \frac{\partial^2}{\partial z^2}u_3 + \sum_{i=1}^{m_a}\Lambda_i^a(z)u_{C,i}^a = 0 \quad \text{for} \quad z \in \{0,1\}. \qquad (4.116)$$

The change in the longitudinal boundary conditions also brings about that the measured scaled electric charge of (4.102), after a single integration by parts, takes the form

$$Q_j^s = \Gamma_j^s\,(1)\,u_1\,(1,t) + \int_0^1\left(-\frac{\partial}{\partial z}\Gamma_j^s\,(z)\,u_1 + \frac{\Gamma_j^s\,(z)}{2}\left(\frac{\partial}{\partial z}u_3\right)^2\right)dz.$$

(4.117)

4.3.7 Controller Design for the Beam with an Axial Support Motion

Comparing (4.115) with (4.117), we can see that one symmetrically supplied piezoelectric actuator and sensor layer couple ($m_s = 1$) with the shaping function $\Lambda_1^s\,(z) = \Gamma_1^s\,(z) = 1$ together with the control law

$$u_{C,1}^s = Q_1^s \qquad (4.118)$$

cancels the effect of the axial disturbance $u_1\,(1,t)$ in the sense of the disturbance compensation of Subsection 4.2.6 [62]. Thus, the equations of motion
(4.115) and (4.116) result in

$$\frac{\partial}{\partial t}p + \frac{\partial^4}{\partial z^4}u_3 + \sum_{j=1}^{m_a}\frac{\partial^2}{\partial z^2}\Lambda_j^a(z)u_{C,j}^a = 0 \qquad (4.119)$$

with the boundary conditions

$$u_3 = 0 \quad\text{and}\quad \frac{\partial^2}{\partial z^2}u_3 + \sum_{j=1}^{m_a}\Lambda_j^a(z)u_{C,j}^a = 0 \quad\text{for}\quad z \in \{0,1\}. \qquad (4.120)$$

In order to suppress vibrations due to an initial deflection or a lateral loading,
an additional non-linear H_∞- or PD-controller for the system (4.119) can be
designed. As already mentioned in the Remark 4.7 the two collocated actuator
and sensor layer couples with the shaping functions (4.104) and (4.108) in
combination with the control laws (4.111) or (4.113) will suppress all even
and odd excited deflection modes.

5. Hydraulic Drive Systems

In this chapter, we discuss two types of hydraulic drive systems, namely a valve-controlled translational piston actuator and a pump-displacement-controlled rotational piston actuator, with particular emphasis on the aspect of control. In [13] these two hydraulic drive types are classified as the two basic ways to control the flow of fluid power to a load. Generally, hydraulic actuators are used to convert hydraulic energy to mechanical energy and *vice versa*. The main advantages of hydraulic power transmission are the light weight and the relatively small volume of the hydraulic components. In contrast to this, the whole equipment for a hydraulic power system is rather expensive and power transmission over a longer distance is nearly impossible. In the sense of the spirit of this book, a strong analytic mathematical description of the considered hydraulic drives is also presented here. The model simplifications and the simplifying assumptions for the controller design are always clearly pointed out to the reader. Furthermore, this chapter contains two industrial applications, namely the hydraulic gap control with eccentricity compensation for rolling mills and the swash-plate mechanism of a hydrostatic drive unit.

5.1 Valve-controlled Translational Piston Actuator

Let us consider the basic configuration of a valve-controlled translational piston actuator as it is presented in Fig. 5.1. Here, $V_{0,1}$ and $V_{0,2}$ denote the volumes of the forward and return chamber for $x_k = 0$, A_1 and A_2 are the effective piston areas, x_k is the displacement of the piston, m_k is the sum of the piston mass and all masses rigidly connected to the piston, q_1 is the flow from the valve to the forward chamber, q_2 denotes the flow from the return chamber to the valve, q_{int} is the internal leakage flow and the external leakage flows are $q_{ext,1}$ and $q_{ext,2}$. Since for all subsequent considerations the volumes $V_{0,1}$ and $V_{0,2}$ and the piston areas A_1 and A_2 are constant, but arbitrary, the theory as presented covers all the different configurations of single- and double-ended as well as single- and double-acting hydraulic actuators. For construction details for these configurations see e.g., [13], [96], [101].

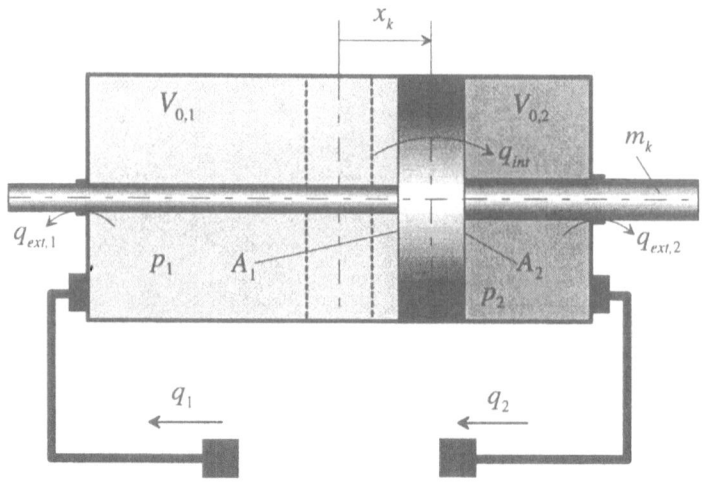

Fig. 5.1. Schematic diagram of a translational hydraulic piston actuator.

5.1.1 Mathematical Model

Before starting with the derivation of the mathematical model for the valve-controlled translational piston actuator of Fig. 5.1, we have to define the constitutive law of the liquid, in our case oil. It is well known that the mass density of oil ρ_{oil} changes with both pressure p and temperature T. To start with, we add the hypothesis that the density is independent of the temperature. Furthermore, since changes in density due to pressure are relatively small, it is usual to use a linearized constitutive law. However, in the literature one can find various definitions for the so-called isothermal bulk modulus β_T of oil. Here, we follow the definition of [13], [81] or recently [88]

$$\frac{1}{\beta_T} = -\frac{1}{V}\left(\frac{\partial V}{\partial p}\right)_{T\,=\,\text{const.}} \tag{5.1}$$

with the total volume V and the pressure p. If the mass in the considered volume V is assumed to be constant, (5.1) is equivalent to

$$\frac{1}{\beta_T} = \frac{1}{\rho_{oil}}\left(\frac{\partial \rho_{oil}}{\partial p}\right)_{T\,=\,\text{const.}} \tag{5.2}$$

From Fig. 5.1 the continuity equations for the two chambers read as

$$\frac{\mathrm{d}}{\mathrm{d}t}\left(\rho_{oil}\left(p_1\right)\left(V_{0,1} + A_1 x_k\right)\right) = \rho_{oil}\left(p_1\right)\left(q_1 - q_{int} - q_{ext,1}\right)$$
$$\frac{\mathrm{d}}{\mathrm{d}t}\left(\rho_{oil}\left(p_2\right)\left(V_{0,2} - A_2 x_k\right)\right) = \rho_{oil}\left(p_2\right)\left(q_{int} - q_{ext,2} - q_2\right), \tag{5.3}$$

provided that the oil temperature T remains constant and the oil is isotropic. Inserting (5.2) into (5.3) and using the fact that the leakage flows are laminar, we get

$$\frac{\mathrm{d}}{\mathrm{d}t}p_1 = \frac{\beta_T}{(V_{0,1} + A_1 x_k)} \left(q_1 - A_1 v_k - C_{int}\left(p_1 - p_2\right) - C_{ext,1}p_1\right)$$

$$\frac{\mathrm{d}}{\mathrm{d}t}p_2 = \frac{\beta_T}{(V_{0,2} - A_2 x_k)} \left(-q_2 + A_2 v_k + C_{int}\left(p_1 - p_2\right) - C_{ext,2}p_2\right)$$

(5.4)

with $v_k = \mathrm{d}x_k/\mathrm{d}t$ and the leakage coefficients C_{int}, $C_{ext,1}$ and $C_{ext,2}$. Suppose that the servo valve is an ideal critical center valve and that it is rigidly connected to a constant pressure pump. The flows from and to the valve, q_1 and q_2, can then be calculated by

$$q_1 = K_{v,1}\sqrt{p_S - p_1}\,\mathrm{sg}\left(x_v\right) - K_{v,2}\sqrt{p_1 - p_T}\,\mathrm{sg}\left(-x_v\right)$$

$$q_2 = K_{v,2}\sqrt{p_2 - p_T}\,\mathrm{sg}\left(x_v\right) - K_{v,1}\sqrt{p_S - p_2}\,\mathrm{sg}\left(-x_v\right)$$

(5.5)

with the supply and the tank pressure p_S and p_T, the valve displacement x_v, the function $\mathrm{sg}\left(x_v\right) = x_v$ for $x_v > 0$ and $\mathrm{sg}\left(x_v\right) = 0$ for $x_v \leq 0$ and the coefficients $K_{v,i} = C_d A_{v,i}\sqrt{2/\rho_{oil}}$, $i = 1, 2$, where $A_{v,i}$ is the orifice area and C_d is the discharge coefficient (see, e.g., [96], [101]). Often the dynamics of the servo valve are much faster than the other components of the hydraulic adjustment system, and therefore, we will ignore the servo valve dynamics and consider the valve displacement x_v to be the plant input to the system.

For the later considerations let us take a three-land-four-way spool valve as it is shown in Fig. 5.2. After the spool valve has been in operation for

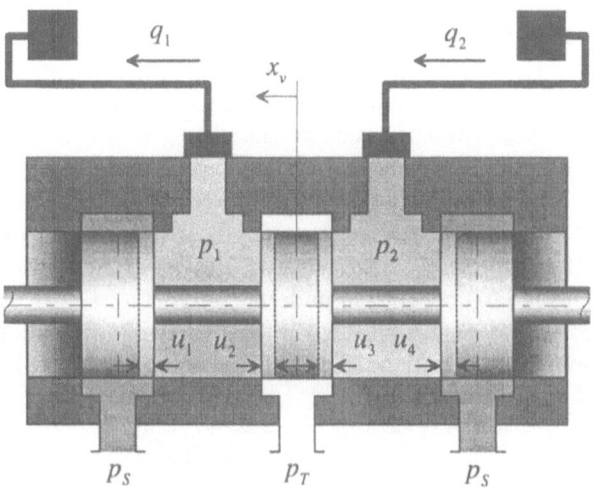

Fig. 5.2. Schematic diagram of a three-land-four-way spool valve.

a while, the orifice edges of a critical center valve are eroded by abrasive material in the oil and hence the leakage flows of the valve are increasing. If such a worn critical center valve is in its centered position, it behaves somewhat like an open center valve. Let u_1, u_2, u_3 and u_4 denote the underlap lengths of the four orifice edges in the case when the valve is centered (dotted valve piston in Fig. 5.2), then (5.5) changes to

$$
\begin{aligned}
q_1 &= \hat{K}_{v,1}\sqrt{p_S - p_1}\,\mathrm{sg}\,(x_v + u_1) - \hat{K}_{v,2}\sqrt{p_1 - p_T}\,\mathrm{sg}\,(-x_v + u_2) \\
q_2 &= \check{K}_{v,2}\sqrt{p_2 - p_T}\,\mathrm{sg}\,(x_v + u_3) - \check{K}_{v,1}\sqrt{p_S - p_2}\,\mathrm{sg}\,(-x_v + u_4)
\end{aligned}
\tag{5.6}
$$

with the modified valve coefficients $\hat{K}_{v,i}$ and $\check{K}_{v,i}$, $i = 1,2$ due to the varying discharge coefficients. The leakage flows of the valve and the piston themselves do not have much influence on the dynamic behavior of the hydraulic adjustment system and therefore, for the purpose of a controller design, they may be neglected. But the leakage flows of the valve may cause pressures p_1 and p_2 in the forward and return chamber, with a considerable offset value p_{off} from symmetrical pressure conditions, and this effect is no longer negligible. In order to clarify this statement, one can easily see that the same piston force F_h can be obtained under totally different pressure conditions

$$
F_h = A_1 p_1 - A_2 p_2 = A_1\,(p_1 + p_{off}) - A_2\left(p_2 + \frac{A_1}{A_2}p_{off}\right)
\tag{5.7}
$$

with p_{off} arbitrary, but restricted within certain boundaries, namely $p_T - p_1 < p_{off} < p_S - p_1$ and $p_T - p_2 < \frac{A_1}{A_2}p_{off} < p_S - p_2$. The problem with this offset pressure is that it causes a different dynamic behavior for positive and negative movements of the piston, because the flows from and to the valve are driven by asymmetric pressure differences. In the closed-loop the abrasion induced truncation of the orifice edges, described by the underlap lengths u_i, $i = 1,\dots,4$, causes also a resulting stationary offset $x_{v,off}$ of the valve position from its original centered position. In general, $x_{v,off}$ depends on the stationary pressures in the two chambers. To give an explanation, let us consider the stationary case where the piston force F_h is held at a predefined value $F_{h,d}$ by an underlying ideal force controller. Then, by ignoring the leakage flows in (5.4), we see that the flows from and to each chamber must be stationary equal, i.e., $q_1 = q_2 = 0$. Thus, for a given worn critical center valve (see (5.6)), with fixed underlap lengths u_i, $i = 1,\dots,4$ and valve coefficients $\hat{K}_{v,i}$ and $\check{K}_{v,i}$, $i = 1,2$, we have three equations for characterizing the stationary values of the chamber pressures $p_{1,s}$, $p_{2,s}$ and of the valve displacement $x_{v,s}$, that are

$$
\begin{aligned}
F_{h,d} &= A_1 p_{1,s} - A_2 p_{2,s} \\
\hat{K}_{v,1}\sqrt{p_S - p_{1,s}}\,\mathrm{sg}\,(x_{v,s} + u_1) &= \hat{K}_{v,2}\sqrt{p_{1,s} - p_T}\,\mathrm{sg}\,(-x_{v,s} + u_2) \\
\check{K}_{v,2}\sqrt{p_{2,s} - p_T}\,\mathrm{sg}\,(x_{v,s} + u_3) &= \check{K}_{v,1}\sqrt{p_S - p_{2,s}}\,\mathrm{sg}\,(-x_{v,s} + u_4)\ .
\end{aligned}
\tag{5.8}
$$

The offset pressure p_{off} and the valve position offset $x_{v,off}$ result directly from the stationary values $p_{1,s}, p_{2,s}$ and $x_{v,s}$. It is quite clear that the underlap lengths u_i, $i = 1, \ldots , 4$ and the valve coefficients $\hat{K}_{v,i}$ and $\check{K}_{v,i}$, $i = 1, 2$ of the worn critical center valve are not known. But for the controller design it suffices to take into account the main sources for the non-linear behavior of the hydraulic adjustment system, that are

- the change of the oil volumes of the chambers, $V_{0,1}$ and $V_{0,2}$, with the piston position x_k, which causes a position dependent stiffness,
- the non-linear dependence of the flows q_1 and q_2 from (5.5) on the chamber pressures p_1, p_2 and the valve displacement x_v,
- the offset pressure p_{off} in the cylinder chambers and
- the stationary valve offset $x_{v,off}$ of the valve position.

By neglecting the leakage flows and inserting (5.4) into $F_h = A_1 p_1 - A_2 p_2$, we get the following differential equation for the piston force

$$\frac{d}{dt} F_h = \frac{\beta_T A_1 (q_1 - A_1 v_k)}{(V_{0,1} + A_1 x_k)} - \frac{\beta_T A_2 (-q_2 + A_2 v_k)}{(V_{0,2} - A_2 x_k)} . \tag{5.9}$$

Since in general the stationary valve offset $x_{v,off}$ as a function of the stationary chamber pressures, $p_{1,s}$ and $p_{2,s}$, is not available, an average value of the valve offset $\bar{x}_{v,off}$ will be used. This value can be obtained mostly within a calibration process. Thus, we modify the relations for the flows from and to the valve, q_1 and q_2, due to (5.5) in the form

$$\begin{aligned} q_1 &= K_{v,1}\sqrt{p_S - p_1}\, \mathrm{sg}\,(\bar{x}_v) - K_{v,2}\sqrt{p_1 - p_T}\, \mathrm{sg}\,(-\bar{x}_v) \\ q_2 &= K_{v,2}\sqrt{p_2 - p_T}\, \mathrm{sg}\,(\bar{x}_v) - K_{v,1}\sqrt{p_S - p_2}\, \mathrm{sg}\,(-\bar{x}_v) \end{aligned} \tag{5.10}$$

with $\bar{x}_v = x_v - \bar{x}_{v,off}$. In fact, the mathematical model described by (5.9) and (5.10) contains all the main non-linearities as mentioned above and therefore, it will subsequently serve as a basis for the controller design.

The equations of motion for the piston are considered to be of the type

$$\begin{aligned} \frac{d}{dt} x_k &= v_k \\ \frac{d}{dt} v_k &= \frac{1}{m_k} (F_h - d_k v_k - F_{load}) \end{aligned} \tag{5.11}$$

with the hydraulic force F_h due to (5.7), the damping coefficient d_k and the external load force F_{load} on the piston, which is assumed to be constant but unknown. For clearer understanding, the terminology used is summarized in Table 5.1.

5.1.2 Controller Design

In the field of control of hydraulic actuators numerous textbooks and many papers have been published in recent years. It is neither within the scope of

Table 5.1. Nomenclature for the valve-controlled translational piston actuator.

A_1, A_2	:	effective piston areas
$A_{v,1}, A_{v,2}$	:	orifice areas
$C_{ext,1}, C_{ext,2}$	:	external laminar leakage coefficients
C_d	:	discharge coefficient
C_{int}	:	internal laminar leakage coefficient
F_{load}	:	external load force
$F_h, (F_{h,d})$	:	hydraulic force (desired)
d_k	:	damping coefficient
$K_{v,i}, \hat{K}_{v,i}, \check{K}_{v,i}, \ i = 1, 2$	:	valve coefficients
m_k	:	sum of rigidly connected piston masses
$p_1, p_2, (p_{1,s}, p_{2,s})$	:	chamber pressures (stationary)
p_{off}	:	offset pressure in the chambers
p_S, p_T	:	supply pressure, tank pressure
q_1	:	flow from the valve to the forward chamber
q_2	:	flow from the return chamber to the valve
$q_{ext,1}, q_{ext,2}$	:	external leakage flows
q_{int}	:	internal leakage flow
T	:	temperature
$u_i, \ i = 1, 2, 3, 4$	:	underlap lengths
$V_{0,1}, V_{0,2}$	:	volumes of the two chambers for $x_k = 0$
v_k, x_k	:	piston velocity, piston position
$x_v, \bar{x}_v, (x_{v,s})$	:	valve displacement (stationary)
$x_{off}, \bar{x}_{off}$	:	offset of the valve displacement
β_T	:	isothermal bulk modulus of oil
ρ_{oil}	:	density of oil

this section nor our intention to go into the details of all these different approaches. The classical methods of hydraulic control are based on a linearized description of the plant around a fixed reference position (see, e.g., [96]). In many practical applications, however, these linear controllers are sufficient in terms of accuracy and dynamic performance and hence are still very common in industry. But as we have seen so far, the hydraulic plant exhibits significant non-linearities and therefore, an increase in the performance of the closed-loop can only be achieved by controllers that take into account the non-linear nature of the system. In the literature, linear controllers either with an adaptation mechanism (e.g., [15], [111], [136]) or robustly designed (e.g., [27]) are often suggested as a means of coping with the non-linearities.

But all these approaches assume a linear nominal model of the hydraulic system and so they mostly suffer from the lack of a stability proof.

The mathematical model (5.9), (5.10) and (5.11) has a very pleasing structure. It is member of a special class of non-linear systems where the control input, in our case x_v, appears affine on the right hand side of the state space representation for $x_v > \bar{x}_{v,off}$ and $x_v \leq \bar{x}_{v,off}$ respectively. This type of non-linear control systems is known as the so-called AI- (affine input) system, see, e.g., [52], [103], [135], [144]. Now, for AI-systems the literature offers a variety of methods for solving the analysis and control synthesis tasks. Without going into the details here, the reader who is familiar with AI-systems can easily see that the mathematical model (5.9), (5.10) and (5.11) with the control input $x_v > \bar{x}_{v,off}$ $(x_v \leq \bar{x}_{v,off})$ is exact input-state linearizable. For an efficient tool for the analysis and synthesis of AI-systems, the reader is referred to the computer algebra package AIsys, which can be obtained free of charge from the webpage of the Maple Application Center [70]. The knowledge that the system is exact input-state linearizable greatly simplifies the non-linear controller design. In the recent literature one can find various papers, which take more or less advantage of this property. See, e.g., [77] for a differential geometric approach or [12] for the application of the theory of flat systems. Another very interesting approach, based on the singular perturbation analysis in combination with exact feedback linearization, especially developed for oscillation drives, is presented in [86]. In fact, it turns out that if the nominal model parameters match with the reality and the measured signals are not very much corrupted by noise, the performance of the closed-loop is excellent throughout the operating range and the deficiencies of the linear control strategies can be overcome. But in some applications, those controllers, which have to rely on the knowledge of the piston velocity, have problems in the case of noisy measurements and/or parameter variations. The parameter variations may typically occur e.g., with a change in the friction, or if the exact value of the bulk modulus of oil β_T in (5.9) differs from the nominal value because of entrapped air and/or mechanical compliance. This is also why some of the proposed non-linear controllers are hardly used in the rough industrial environment.

However, one of the key observations is that the non-linear controller must not contain a velocity signal v_k, because in many applications the velocity v_k can only be obtained by approximate differentiation of the position signal x_k, which is known to be very sensitive to noise. The differentiation process would not cause any problems, if high-precision position sensors in combination with a very high sampling time are used, but this is not the normal situation in the industry. Therefore, following [67], we will subsequently propose a non-linear controller for the hydraulic system (5.9), (5.10) and (5.11), where only the pressures p_1 and p_2 of the two chambers of the hydraulic cylinder and the displacement of the hydraulic piston x_k are assumed to be directly available through measurement. The idea is to perform an exact input-output linearization (see again e.g., [52], [103], [135], [144] for the theoretical background on this topic) for the output

$$z = F_h + \beta_T \ln \frac{(V_{0,1} + A_1 x_k)^{A_1}}{(V_{0,2} - A_2 x_k)^{A_2}} \; . \tag{5.12}$$

Clearly, z is nothing else than the piston force F_h due to the pressures p_1 and p_2 extended by the deviation of the force due to the change of the chamber volumes. One can immediately see by calculating

$$\frac{d}{dt} z = \frac{\beta_T A_1 q_1}{V_{0,1} + A_1 x_k} + \frac{\beta_T A_2 q_2}{V_{0,2} - A_2 x_k} \tag{5.13}$$

that z remains constant as long as the flows from and to the valve, q_1 and q_2, are zero. Summarizing, we can write the mathematical model of the hydraulic system (5.9), (5.10) and (5.11) in the new state variables z, x_k and v_k in the form

$$\frac{d}{dt} z = \frac{\beta_T A_1}{V_{0,1} + A_1 x_k} q_1 + \frac{\beta_T A_2}{V_{0,2} - A_2 x_k} q_2$$

$$\frac{d}{dt} x_k = v_k$$

$$\frac{d}{dt} v_k = \frac{1}{m_k} \left(z - d_k v_k - \beta_T \ln \frac{(V_{0,1} + A_1 x_k)^{A_1}}{(V_{0,2} - A_2 x_k)^{A_2}} - F_{load} \right) \tag{5.14}$$

with q_1 and q_2 from (5.10).

Proposition 5.1. *Given a hydraulic system described by (5.14). Let us assume that the position $x_{k,ref}$ and the unknown, but constant load force F_{load} determine an admissible stationary point of the system. Then the control law*

$$x_v = \bar{x}_{v,off} + \left(\frac{A_1}{V_{0,1} + A_1 x_k} K_{v,1} \sqrt{p_S - p_1} + \right.$$

$$\left. \frac{A_2}{V_{0,2} - A_2 x_k} K_{v,2} \sqrt{p_2 - p_T} \right)^{-1} v \tag{5.15}$$

for $x_v > \bar{x}_{v,off}$ and

$$x_v = \bar{x}_{v,off} - \left(\frac{A_1}{V_{0,1} + A_1 x_k} K_{v,2} \sqrt{p_1 - p_T} + \right.$$

$$\left. \frac{A_2}{V_{0,2} - A_2 x_k} K_{v,1} \sqrt{p_S - p_2} \right)^{-1} v \tag{5.16}$$

for $x_v \leq \bar{x}_{v,off}$ with

$$v = \alpha_1 \ln \left(\left(\frac{V_{0,1} + A_1 x_{k,ref}}{V_{0,1} + A_1 x_k} \right)^{A_1} \left(\frac{V_{0,2} - A_2 x_k}{V_{0,2} - A_2 x_{k,ref}} \right)^{A_2} \right) \tag{5.17}$$

guarantees that the stationary point is asymptotically stable throughout the operating domain for all possible values of the bulk modulus of oil $\beta_T >$ $\beta_{T,\min}$, provided that the inequality conditions

$$0 < \alpha_1 < \min \left(\frac{d_k}{m_k}, \frac{c_k \beta_{T,\min}}{d_k} \right) \tag{5.18}$$

with

$$c_k = \frac{A_1 A_2 \left(\sqrt{A_1} + \sqrt{A_2} \right)^2}{(A_1 V_{0,2} + A_2 V_{0,1})} \tag{5.19}$$

are satisfied. By means of the control parameter α_1 the dynamics of the closed-loop can easily be adjusted.

Proof. Substituting the control law (5.15) and (5.16) into (5.10) and afterwards into (5.14), we obtain the closed-loop system written in deviations Δ around the stationary point, determined by $x_{k,ref}$ and F_{load}, by

$$\frac{\mathrm{d}}{\mathrm{d}t} \Delta z = -\alpha_1 f\left(\Delta x_k\right)$$
$$\frac{\mathrm{d}}{\mathrm{d}t} \Delta x_k = \Delta v_k \tag{5.20}$$
$$\frac{\mathrm{d}}{\mathrm{d}t} \Delta v_k = \frac{1}{m_k} \left(\Delta z - d_k \Delta v_k - f\left(\Delta x_k\right) \right)$$

with

$$f\left(\Delta x_k\right) = \beta_T \ln \left(\left(\frac{V_{0,1} + A_1 \left(\Delta x_k + x_{k,ref}\right)}{V_{0,1} + A_1 x_{k,ref}} \right)^{A_1} \right.$$
$$\left. \left(\frac{V_{0,2} - A_2 x_{k,ref}}{V_{0,2} - A_2 \left(\Delta x_k + x_{k,ref}\right)} \right)^{A_2} \right) . \tag{5.21}$$

One of the key observations here is that the non-linear function $f\left(\Delta x_k\right)$ satisfies the sector condition

$$0 \leq \beta_T c_k \Delta x_k^2 \leq f\left(\Delta x_k\right) \Delta x_k < \infty \tag{5.22}$$

with

$$c_k = \frac{A_1 A_2 \left(\sqrt{A_1} + \sqrt{A_2} \right)^2}{(A_1 V_{0,2} + A_2 V_{0,1})} \tag{5.23}$$

for $\Delta x_{k,\min} < \Delta x_k < \Delta x_{k,\max}$ with $\Delta x_{k,\min} = -V_{0,1}/A_1 - x_{k,ref}$ and $\Delta x_{k,\max}$ $= V_{0,2}/A_2 - x_{k,ref}$.

Now, the mathematical model (5.20) and (5.21) can be represented as a feedback interconnection of a reachable and observable linear subsystem with the transfer function

$$Z(s) = \frac{s + \alpha_1}{m_k s^3 + d_k s^2 + \beta_T c_k s + \beta_T c_k \alpha_1} \tag{5.24}$$

and a static non-linearity

$$\psi(\Delta x_k) = f(\Delta x_k) - \beta_T c_k \Delta x_k \tag{5.25}$$

with $f(\Delta x_k)$ from (5.21). Compare this with Fig. 1.6 of Chapter 1. It can be immediately seen that the transfer function $Z(s)$ is Hurwitz, if and only if for α_1 the condition

$$0 < \alpha_1 < \frac{d_k}{m_k} \tag{5.26}$$

holds. Thus, we may apply the well known Popov criterion (see Theorem 1.7 and Remark 1.12 of Chapter 1 or e.g., [59], [144]), which says that the system (5.24) with the non-linearity (5.25) satisfying the sector condition

$$0 \le \psi(\Delta x_k) \Delta x_k < \infty \tag{5.27}$$

for $\Delta x_{k,\min} < \Delta x_k < \Delta x_{k,\max}$ with $\Delta x_{k,\min} = -V_{0,1}/A_1 - x_{k,ref}$ and $\Delta x_{k,\max}$ $= V_{0,2}/A_2 - x_{k,ref}$ is absolutely stable if there exists an $\eta \ge 0$, such that

$$\mathrm{Re}\left(Z(j\omega)\right) - \eta\omega\,\mathrm{Im}\left(Z(j\omega)\right) > 0 \tag{5.28}$$

for all $\omega \in R$. Choosing $\eta = m_k/(d_k - \alpha_1 m_k) > 0$ and performing some computer algebra calculations, we obtain the result that (5.28) is fulfilled for

$$\alpha_1 < \frac{c_k \beta_T}{d_k} \ . \tag{5.29}$$

The inequality conditions (5.26) and (5.29) can be combined to

$$0 < \alpha_1 < \min\left(\frac{d_k}{m_k}, \frac{c_k \beta_{T,\min}}{d_k}\right) \tag{5.30}$$

with $\beta_{T,\min}$ as the lower bound for the bulk modulus of oil β_T. Note that α_1 can always be chosen such that $-1/\eta$ is not a pole of $Z(s)$ from (5.24).

At first sight this result seems to be a local one because the sector condition (5.27) holds only in a finite domain $\Delta x_{k,\min} < \Delta x_k < \Delta x_{k,\max}$. But in the next step we will show that the set $\Omega = \{\Delta z, \Delta x_k, \Delta v_k \in R | \Delta x_{k,\min} < \Delta x_k < \Delta x_{k,\max}\}$ is positively invariant, that means, every trajectory starting in Ω stays for all future time in Ω. Therefore, we can deduce that every admissible stationary point defined by $x_{k,ref}$ and F_{load} is asymptotically stable in Ω. For this purpose let us rewrite (5.20) in the original state variables

$$\frac{\mathrm{d}}{\mathrm{d}t}\Delta F_h = -\alpha_1 f(\Delta x_k) - \frac{\partial}{\partial \Delta x_k} f(\Delta x_k)\,\Delta v_k$$

$$\frac{\mathrm{d}}{\mathrm{d}t}\Delta x_k = \Delta v_k \tag{5.31}$$

$$\frac{\mathrm{d}}{\mathrm{d}t}\Delta v_k = \frac{1}{m_k}(\Delta F_h - d_k \Delta v_k)\ .$$

Investigating the limit $\Delta x_k \to \Delta x_{k,\max}$, we obtain $\lim_{\Delta x_k \to \Delta x_{k,\max}} f(\Delta x_k) = \infty$. Since now in (5.31) $|\mathrm{d}\Delta F_h/\mathrm{d}t| \to \infty$, the system (5.31) can be decomposed in a fast and in a slow manifold [59]. The trajectories in the fast manifold rapidly descend to the manifold

$$\Delta v_k = -\frac{\alpha_1 f(\Delta x_k)}{\dfrac{\partial}{\partial \Delta x_k} f(\Delta x_k)} \tag{5.32}$$

and since $\Delta x_k \to \Delta x_{k,\max}$ it follows that $\Delta v_k < 0$. The same holds for $\Delta x_k \to \Delta x_{k,\min}$ then $\Delta v_k > 0$. So we see that the trajectories do not cross the boundary of Ω. And this completes the proof. ∎

5.2 Application: Hydraulic Gap Control (HGC) in Rolling Mills

5.2.1 System Description

Fig. 5.3 presents the schematic diagram of a four-high mill stand with the hydraulic adjustment system acting on the upper backup roll. The work rolls are effectively used for the strip deformation whereas the backup rolls serve to support the work rolls in order to prevent them from bending too much. The rolls are running in so-called chocks, which can move vertically in the mill housing to enable a change of the roll gap. The thickness of the rolled strip is predominantly determined by the gap between the two work rolls which is initially set by a pass-line adjustment mechanism. The hydraulic positioning system is then used for an exact and fast-acting position control. This is necessary since the required tolerances of the final strip product are very tight. Subsequently, we assume without restriction of generality that the hydraulic adjustment system consists of either a single- or a double-acting hydraulic ram which can be described by a mathematical model of the form (5.9) and (5.10).

5.2.2 Mill Stand Model

In general, the mathematical models of the mill stand used by the rolling mill's design engineers are highly complex and based on a finite element calculation. The problem is that these models are not useful for the purpose of designing a controller. Depending on the application, one can find various simpler dynamic models in the literature, which are composed of discrete masses, springs and dampers. So, for instance, in [3] the stand model was specifically derived for the identification of the mill stretch coefficient and the deformation resistance. A simple mill stand model of this category, where the effect of the roll eccentricities of the work and backup rolls and the friction between the chocks and the mill housing are also taken into account, is depicted in Fig. 5.4. The mill stand is modeled in the form of the discrete masses m_1 (mill housing), m_2 (upper backup roll + chock + piston), m_3 (upper work roll + chock), the dampers d_1, d_2 and the springs c_1, c_2 with the constant lengths l_1, l_2 and l_3. The pass-line is ideally supposed to be kept at a constant position and to coincide with the inertial frame.

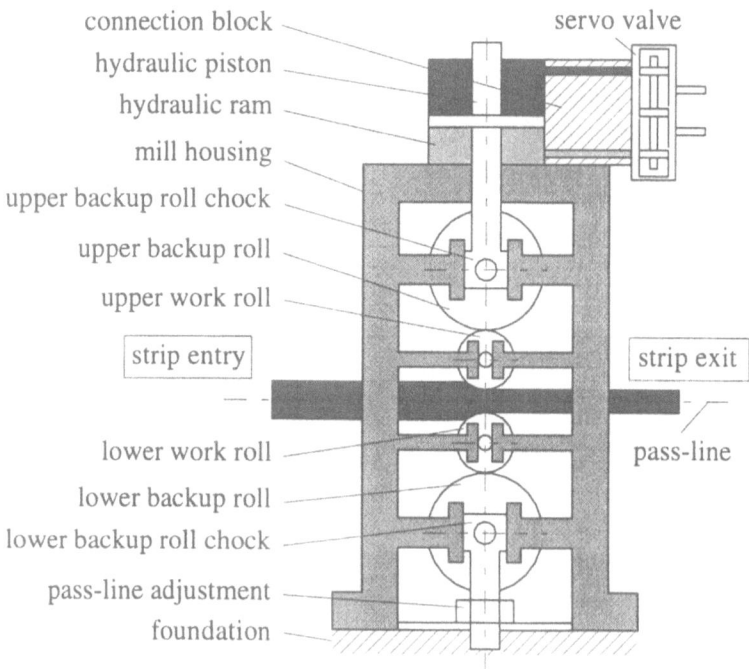

Fig. 5.3. Schematic diagram of a four-high mill stand.

Then the equations of motion take the following form

$$\frac{d}{dt}x_i = v_i \quad , \quad i = 1, \dots, 3$$

$$m_1\frac{d}{dt}v_1 = -c_1(x_1 - l_1) - d_1 v_1 + F_h - m_1 g - F_{fric,1} - F_{fric,2}$$

$$m_2\frac{d}{dt}v_2 = -F_h - m_2 g + F_{fric,1} - d_2(v_2 - v_3 - v_{e,B} + v_{e,W}) -$$
$$c_2(x_2 - x_3 - l_2 + l_3 - x_{e,B} + x_{e,W})$$

$$m_3\frac{d}{dt}v_3 = F_r - m_3 g + F_{fric,2} + d_2(v_2 - v_3 - v_{e,B} + v_{e,W}) +$$
$$c_2(x_2 - x_3 - l_2 + l_3 - x_{e,B} + x_{e,W})$$

$$\frac{d}{dt}x_{e,B} = v_{e,B}$$

$$\frac{d}{dt}x_{e,W} = v_{e,W}$$

(5.33)

with the gravitational constant g, the piston force F_h due to (5.7), the roll force F_r, the friction forces $F_{fric,1}$ and $F_{fric,2}$ between the work and backup roll chocks and the mill housing and the axial deviations $x_{e,W}(t)$ and $x_{e,B}(t)$ between the roll barrel and the roll neck of the work and backup rolls, respectively. These axial deviations may arise for different reasons, such as, inexact roll grinding, non-uniform thermal expansion of the rolls, roll wear or irregularities in the roll bearings. It is important to take this effect into consideration because these disturbances appear as periodic deviations in the strip exit thickness and they are known as roll eccentricities. The reader is encouraged to consult [35] for further details of causes and effects of roll ec-

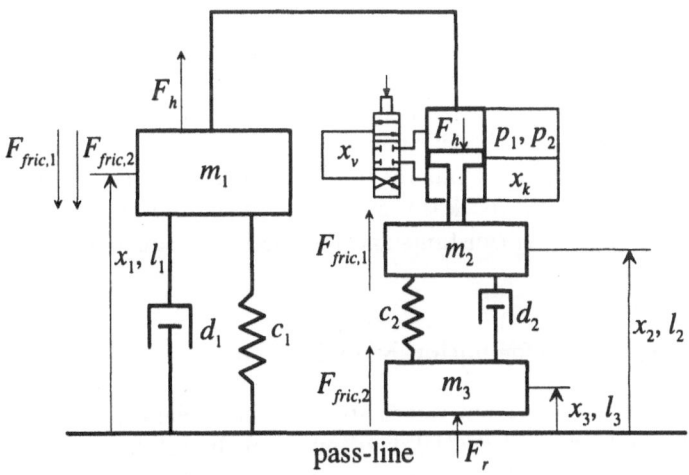

Fig. 5.4. Model of the four-high mill stand.

centricity. For the friction forces $F_{fric,1}$ and $F_{fric,2}$ a special bristle-based dynamic model, the so-called LuGre model, which captures most of the friction phenomena, is used. Following [23], the friction forces $F_{fric,i}$, $i = 1, 2$ read as

$$F_{fric,i} = \sigma_{0,i} z_i + \sigma_{1,i} \frac{\mathrm{d}}{\mathrm{dt}} z_i + \sigma_{2,i} (v_1 - v_{1+i}) \tag{5.34}$$

where z_i represents the average deflection of the bristles with

$$\frac{\mathrm{d}}{\mathrm{dt}} z_i = (v_1 - v_{1+i}) - \frac{\sigma_{0,i}}{\chi} \operatorname{abs}(v_1 - v_{1+i}) z_i \tag{5.35}$$

and

$$\chi = F_{C,i} + (F_{S,i} - F_{C,i}) \exp\left(-\left(\frac{v_1 - v_{1+i}}{v_{s,i}}\right)^2\right). \tag{5.36}$$

Here $F_{C,i}$ denotes the Coulomb friction level, $F_{S,i}$ is the stiction force level, $v_{s,i}$ is the Stribeck velocity and the coefficients $\sigma_{0,i}$, $\sigma_{1,i}$ and $\sigma_{2,i}$ allow us to parametrize the friction model. As is shown in [107], the friction model (5.34), (5.35) and (5.36) has the pleasing property of defining a passive operator from the relative velocity $(v_1 - v_{1+i})$ to the friction force $F_{fric,i}$, if and only if the following condition

$$\sigma_{2,i} - \sigma_{1,i} \left(\frac{F_{S,i}}{F_{C,i}} - 1\right) > 0 \tag{5.37}$$

is satisfied. The strip exit thickness h_{ex} and the displacement of the hydraulic piston x_k, as functions of the state variables of the mill stand, are given by

$$h_{ex} = x_3 - l_3 + x_{e,W} \quad \text{and} \quad x_k = x_1 - l_1 - x_2 + l_2. \tag{5.38}$$

In fact, the mill stand model of Fig. 5.4 allows us to investigate all the essential dynamic effects in a qualitative manner and therefore, it will subsequently be implemented in the mill simulator to test the feasibility of the proposed control concepts. The drawback is that most of the parameters like the damping coefficients, the spring constants and the friction parameters are only known rather inaccurately.

5.2.3 Material Deformation Model

Next, we want to say a few words about the material deformation model of the strip. As long as no spatial distribution of the roll load and no dynamic effects of the deformation process are taken into account, the deformation model can be reduced to a system of implicit non-linear equations of the form

$$f_{roll}\left(F_r,\, h_{ex},\, h_{en},\, \sigma_{ex},\, \sigma_{en},\, \omega_{roll},\, T_{st}\right) = 0 \qquad (5.39)$$

with the roll force F_r, the exit and entry thickness h_{ex} and h_{en}, the specific exit and entry tension σ_{ex} and σ_{en}, the angular velocity of the work or backup roll ω_{roll} and the strip temperature T_{st}. However, the setting up of these deformation models consists of solving various differential and integral equations as well as the definitions of many constitutive parameters, like the friction coefficient μ between the work rolls and the strip and the yield stress σ_F as a function of the strip reduction. We do not intend to go into the details here, but the interested reader is referred to the literature, e.g., [14], [30], [31], [41] or a more recent contribution [29].

5.2.4 HGC with a Double-acting Hydraulic Ram

In most cases, the conventional strip thickness control concepts in rolling mills are based on a cascaded structure with several linear SISO- (single-input single-output) controllers. In the case when the roll gap is adjusted by means of a hydraulic positioning system, the innermost control loop controls the hydraulic piston position and is often referred to as the hydraulic gap control (HGC). It is well known that the underlying physical structure of a rolling mill is a highly complex non-linear coupled process. Since in the classical linear SISO-control approach the inherent non-linearities and the coupling effects are not considered, the performance of the overall closed-loop system is not always satisfactory. However, the literature contains many successful applications of linear MIMO- (multi-input multi-output) controllers to multi-stand rolling mills, which, in fact, overcome the deficiencies of the classical single-loop control concepts (see, e.g., [38], [39], [102]). Nevertheless, all these approaches assume that the process can be described by a linear nominal MIMO-model and the non-linearities of the process are only taken into account by means of uncertainty models. These are supposed to satisfy certain conditions depending on the control design strategy used. This is why in the literature, the proposed controllers are based either on a linear robust approach, e.g., the linear multivariable H_∞-design (see, e.g., [39] and the references cited therein), or on linear self-tuning concepts as presented e.g., in [24]. In many situations the assumptions underlying these models pose no essential restriction, particularly, if the rolling mill is operating around a predefined pass schedule. But if the operating point is changing in a wider range, then the essential non-linearities of the plant to be controlled can no longer be neglected.

Let us assume that the hydraulic adjustment system consists either of a single- or a double-acting hydraulic ram which can be described by a mathematical model of the form (5.9) and (5.10). It is quite apparent that the HGC as the innermost control loop is an essential part of the thickness control concept. We should be therefore aware of the fact that the outer control loops cannot make up for performance deficiencies in the HGC. Thus, we intend to design an HGC such that the following conditions hold:

- The stability of the closed-loop HGC is guaranteed throughout the whole operating range,
- the dynamic behavior remains the same regardless of the initial piston position and
- the dynamic behavior does not change with the aging of the system (offsets in the chamber pressures and the valve position or variation of the exact value of the bulk modulus β_T because of entrapped air, see Subsection 5.1.1 for details).

For a successful practical implementation of the control concept in a rough industrial environment we additionally make the following demands on the HGC:

- Only the pressures p_1, p_2 of the two chambers of the hydraulic cylinder and the displacement of the hydraulic piston x_k are directly measurable,
- the controller should be insensitive to transducer and quantization noise,
- the HGC should cope with parameter inaccuracies in e.g., the damping coefficients, the bulk modulus (entrapped air, mechanical compliance), the friction parameters etc.,
- it should be easy for the commissioning engineer and the maintenance staff to adjust the dynamics of the HGC,
- the HGC must be implemented on a hardware platform with a predefined sampling time and
- the HGC should fit the conventional thickness control concepts for AGC (Automatic Gauge Control)-systems (see, e.g., [24], [112]).

In order to motivate the non-linear controller design for the HGC, we first show in Fig. 5.5 the simulation results of a conventional HGC with a simple P-(proportional) controller, usually used in industry, for four different cases of offset pressures p_{off} and initial piston positions $x_{k,0}$, that are

> **case A:** $p_{off} = 70 \cdot 10^5$ Pa , $x_{k,0} = 90 \cdot 10^{-3}$ m (bottom edge)
> **case B:** $p_{off} = 0 \cdot 10^5$ Pa , $x_{k,0} = 90 \cdot 10^{-3}$ m (bottom edge)
> **case C:** $p_{off} = 70 \cdot 10^5$ Pa , $x_{k,0} = 0 \cdot 10^{-3}$ m (middle)
> **case D:** $p_{off} = 70 \cdot 10^5$ Pa , $x_{k,0} = -90 \cdot 10^{-3}$ m (top edge)

For all four cases the deviation of the piston position $x_k - x_{k,0}$ and the spool valve displacement x_v for a tracking signal $x_{k,ref} = x_{k,0} + 100 \cdot 10^{-6}\sigma\,(t - 0.5) - 100 \cdot 10^{-6}\sigma\,(t - 1)$ in m and for a step disturbance in the entry thickness of $400 \cdot 10^{-6}$ m after 1.5 s at a nominal entry thickness of $3.3 \cdot 10^{-3}$ m is presented. However, one can easily see that the closed-loop dynamics are varying over a wide range and is even different, though the same offset pressure p_{off} and initial piston position $x_{k,0}$, for positive and negative

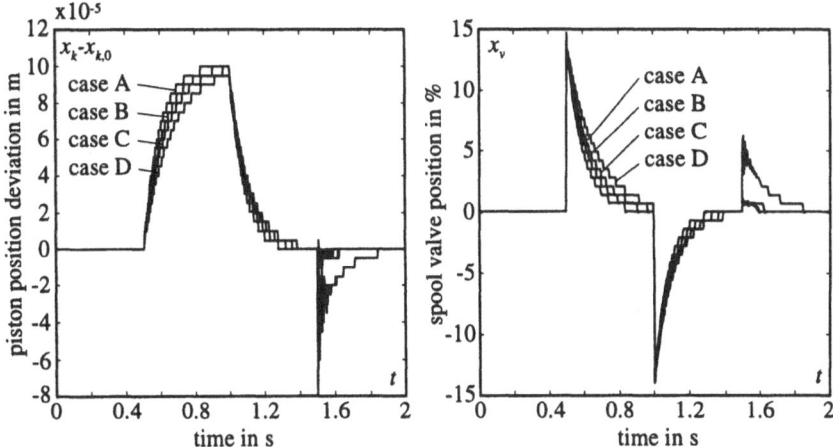

Fig. 5.5. Simulation results of HGC with a conventional linear P-controller: Deviation of the piston position $x_k - x_{k,0}$ and the valve displacement x_v.

reference steps. These simulation results are performed for a hydraulic adjustment system consisting of a double-acting hydraulic ram with the effective piston areas $A_1 = A_2 = 0.68$ m^2, the volumes $V_{0,1} = V_{0,2} = 0.072$ m^3 and a three-land-four-way spool valve with a rated flow of 150 l/min. The supply pressure $p_S = 270 \cdot 10^5$ Pa and the tank pressure $p_T = 0$ Pa. The parameters of the mill stand (5.33) were extracted from a finite element model and the roll eccentricities of the work and backup rolls $x_{e,W}$ and $x_{e,B}$ are set to zero. The friction coefficients for the LuGre model (5.34), (5.35) and (5.36) can be partly determined from measurements and partly from fitting the simulation results the reality. For the deformation process the model of [14], [30] and [31] for cold rolling is used, where the friction coefficient is fixed as $\mu = 0.05$ and the average yield stress $\bar{\sigma}_F = 567 \cdot 10^6$ Nm^{-2}. In order to get realistic simulation results, a quantization of the piston position of $5 \cdot 10^{-6}$ m is included in the simulator and the transducer noise for the pressures in the chambers is modeled as a band-limited white noise.

In some applications the simple P-controller is extended by a servo compensation. Depending on the quality of the servo compensation the influence of the offset pressure p_{off} on the dynamics of the closed-loop can be more or less decreased, but the dependence on the initial piston position $x_{k,0}$ still remains. For the non-linear controller design of the HGC the dynamics of the mill stand and the mill stretch effect will be neglected and hence we may directly apply Proposition 5.1 (see also Remark 5.1). For the purpose of comparability Fig. 5.6 demonstrates the simulation results for the closed-loop HGC with the non-linear controller (5.15) - (5.17) for the identical data, reference and disturbance inputs as in Fig. 5.5. The control parameter α_1 of

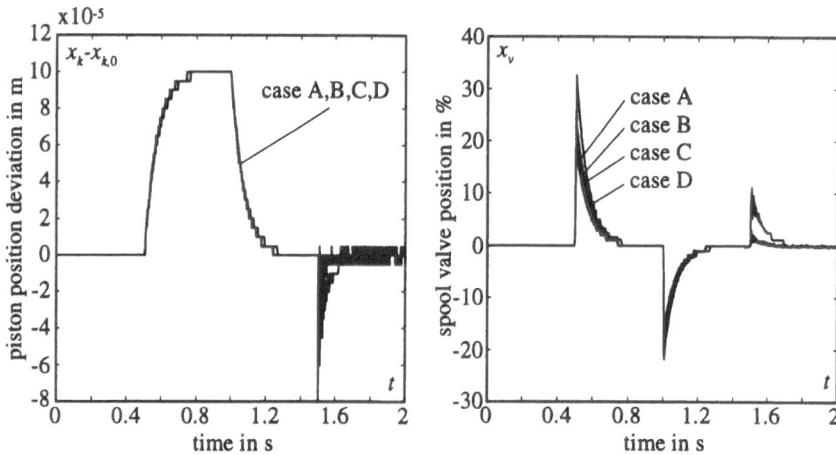

Fig. 5.6. Simulation results of HGC with non-linear controller: Deviation of the piston position $x_k - x_{k,0}$ and the valve displacement x_v.

(5.17) was chosen in such a way that the step responses for the **case A** of the linear and the non-linear controller show a good correspondence. One can easily see that with the non-linear HGC the dynamics of the closed-loop are pretty well the same for all different cases. Further simulation studies allow us to conclude that the non-linear controller has the ability to cope with the essential non-linearities of the system as well as with aging-induced changes of the dynamic behavior. These properties ensure that the outer control loops in a cascaded system can rely on the fact that the dynamics of the inner control loop always remain the same. This prevents unexpected problems, such as unwanted vibrations induced by the piston if it goes to one of the edges of the cylinder: a common problem in industrial operations.

5.2.5 HGC with a Single-acting Hydraulic Ram

Apart from simulation results field tests were also performed at a reversing hot strip mill in the Czech Republic by VOEST-ALPINE Industrieanlagen-bau GmbH, but this time for a single-acting piston configuration (see [104] for details of the plant). In order to get an impression of the size of this mill, we will give a brief description of the hydraulic adjustment system. The head side of the single-acting piston is connected via rigid steel pipes with two three-stage servo-valves. Because the piston is so large, with an effective piston area of $1.13\,\text{m}^2$ and a maximum piston displacement is $0.07\,\text{m}$, two valves with a rated flow of $800\,\text{l/min}$ are driven synchronously to make the required oil flow available. The rod side of the piston is filled with nitrogen at a constant pressure of $3 \cdot 10^5\,\text{Pa}$ and the volume of the connection lines is $0.02\,\text{m}^3$. Without going into the details here the non-linear controller of Proposition 5.1 can also be applied to single-acting piston configurations with some slight but rather easy modifications (see [104]). Fig. 5.7 depicts the measured deviation of the piston displacement and the associated servo-valve position for reference step inputs of $50 \cdot 10^{-6}$ m around an operating point of approximately $7.8 \cdot 10^{-3}$ m and no load in the roll gap, with a traditional linear controller and with the non-linear controller of Proposition 5.1. To enable comparison with the results of the linear controller, and for testing the controller under extreme dynamic situations, the parameter α_1 of the non-linear controller of Proposition 5.1 was adjusted in such a way that the step responses of Fig. 5.7 show an overshoot of approximately 15%. As one can immediately see, in contrast to the non-linear control concept, the traditional linear controller has a different dynamic behavior for steps in the positive and the negative direction. This fact may, particularly limit the achievable thickness tolerances and it badly influences the dynamics of the outer control loops in a cascaded thickness control concept. Of course, for the nominal operation of the plant α_1 is decreased such that the overshoot of the step response vanishes.

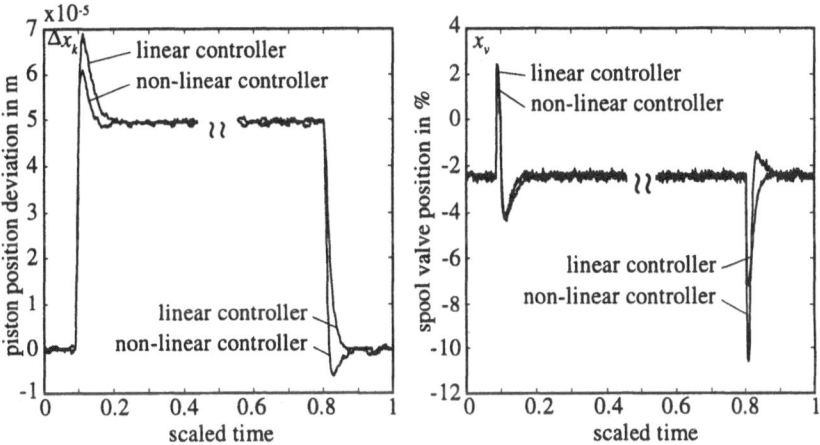

Fig. 5.7. Comparison results of the step responses of the closed-loop HGC with a traditional linear controller and with the non-linear controller of Proposition 5.1 for a single-acting piston configuration measured in a reversing hot strip mill in the Czech Republic by VOEST-ALPINE Industrieanlagenbau GmbH.

Remark 5.1. Ignoring the dynamics of the mill stand (5.33) for the controller design of the HGC is no restriction of the control concept. However, the stability proof of Proposition 5.1 holds with slight modifications, if additionally the mill stretch effect of the stand frame and the roll stack has to be considered. Thereby, it suffices to model the upper rolls, the chocks and the piston as a single rigid mass m_m. The mill stretch effect is taken into account by means of a quasi-stationary measured mill stretch calibration curve $x_{k,str} = f_{str}(F_h)$ [110]. Generally, $f_{str}(F_h)$ can be fairly closely approximated by a linear function $f_{str}(F_h) = F_h/c_{str}$ with c_{str} as the so-called mill stretch coefficient. Figure 5.8 shows the scheme of the simple mill stand

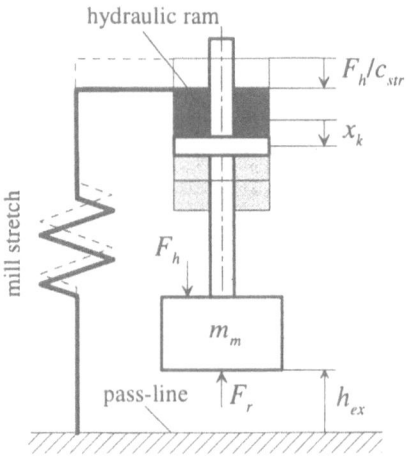

Fig. 5.8. Simple mill stand model with a schematic representation of the mill stretch.

model and the associated equations of motion read as

$$\frac{\mathrm{d}}{\mathrm{d}t}h_{ex} = v_{ex}$$
$$m_m\frac{\mathrm{d}}{\mathrm{d}t}v_{ex} = -F_h - d_m v_{ex} + F_r - m_m g \tag{5.40}$$

with the strip exit thickness h_{ex}, the piston force F_h, the damping coefficient d_m, the gravitational constant g, the total mass of all moving parts m_m and the roll force F_r. In the operating range of practical interest, the roll force F_r can be fairly closely approximated by $F_r = F_{r,0} - c_m h_{ex}$ with the parameter $F_{r,0} > 0$ and the so-called material deformation resistance $c_m > 0$. The displacement of the piston x_k is then given by

$$x_k = \frac{F_h}{c_{str}} - h_{ex} . \tag{5.41}$$

Consequently, the mathematical model (5.14) must be replaced by

$$\frac{d}{dt}\left(F_h + \beta_T \ln \frac{(V_{0,1} + A_1 x_k)^{A_1}}{(V_{0,2} - A_2 x_k)^{A_2}}\right) = \frac{\beta_T A_1}{V_{0,1} + A_1 x_k} q_1 + \frac{\beta_T A_2}{V_{0,2} - A_2 x_k} q_2$$

$$\frac{d}{dt} h_{ex} = v_{ex}$$

$$m_m \frac{d}{dt} v_{ex} = -F_h - d_m v_{ex} + F_{r,0} - c_m h_{ex} - m_m g$$

$$(5.42)$$

with q_1 and q_2 from (5.10) and x_k from (5.41). Substituting the control law (5.15) and (5.16) into (5.10) and afterwards into (5.42), we obtain the closed-loop system written in deviations Δ from the stationary point, determined by $x_{k,ref}$ and $F_{r,0}$, by

$$\frac{d}{dt} \Delta F_h = -\alpha_1 f(\Delta x_k) - \frac{d}{dt} f(\Delta x_k)$$

$$\frac{d}{dt} \Delta h_{ex} = \Delta v_{ex} \qquad (5.43)$$

$$\frac{d}{dt} \Delta v_{ex} = \frac{1}{m_m} \left(-\Delta F_h - d_m \Delta v_{ex} - c_m \Delta h_{ex}\right)$$

with $f(\Delta x_k)$ from (5.21) satisfying the sector condition (5.22) and $\Delta x_k = \Delta F_h / c_{str} - \Delta h_{ex}$. The system (5.43) can now again be decomposed as a feedback interconnection of a reachable and observable linear subsystem with the transfer function $Z(s) = \frac{n_Z(s)}{d_Z(s)}$,

$$n_Z(s) = (s + \alpha_1)\left(m_m s^2 + d_m s + c_m + c_{str}\right)$$
$$d_Z(s) = m_m \left(c_{str} + \beta_T c_k\right) s^3 + \left(\beta_T c_k m_m \alpha_1 + (c_{str} + \beta_T c_k) d_m\right) s^2 +$$
$$\left(\beta_T c_k \left(d_m \alpha_1 + c_m + c_{str}\right) + c_{str} c_m\right) s + \beta_T c_k \alpha_1 \left(c_m + c_{str}\right),$$

$$(5.44)$$

which is for a suitable α_1, $0 < \alpha_1 < \alpha_{1,\max}$, BIBO-stable and a static non-linearity

$$\psi(\Delta x_k) = f(\Delta x_k) - \beta_T c_k \Delta x_k \qquad (5.45)$$

with $f(\Delta x_k)$ from (5.21) and c_k from (5.23). On mild conditions concerning the system parameters the exponential stability can be shown by means of the circle criterion [59], [144]. To obtain these results some tedious lengthy computer algebra calculations have to be performed and we do not intend to demonstrate them here in detail.

Remark 5.2. The non-linear HGC so presented also serves as a key element for new non-linear MIMO-thickness control concepts in rolling mills which follow the trend of modern rolling mills towards tighter thickness tolerances, thinner final strip thicknesses, faster production rates and shorter off-gauge lengths, see, e.g., [74].

5.3 Rejection of Periodic Disturbances (Eccentricity Compensation)

As already mentioned, there is a strong tendency in rolling mills to improve the quality of the rolled product, especially in relation to the thickness tolerances. One problem that may define the limit of the achievable thickness tolerances and which is not eliminated by means of the conventional thickness control concepts like the HGC, is periodic disturbances in the strip exit thickness. Here the frequencies are proportional to the angular velocity of the work and backup rolls, respectively. As was mentioned in Section 5.2.1, these disturbances are said to be caused by so-called roll eccentricities. Typical values of the contribution of the roll eccentricity in the final thickness in hot rolling mills can be about $40 \cdot 10^{-6}$ m and more. The roll eccentricities arise from axial deviations between the roll barrel and the roll necks, due to irregularities in the mill rolls and/or roll bearings [35]. Several patents and papers have been published concerning the active and passive compensation of these eccentricity-induced disturbances. An excellent overview of the results along the more application oriented branch can be found in [35]. Passive compensation methods try only to avoid the gain effect of the roll eccentricity in a mill stretch compensation loop (AGC). Whereas an active eccentricity compensation generates a supplementary signal in the position or force control loop of the hydraulic adjustment system in order to suppress the periodic disturbances in the strip. Some of the proposed active approaches operate in two stages. First, the contribution of the eccentricity is identified directly from the measured force and/or thickness by using FFT (Fast Fourier Transform)-based algorithms, neural networks or least squares methods. Secondly, the resulting amplitude and phase of the eccentricity signal are then fed to a PI (Proportional Integral)-controller or are used for disturbance feedforward compensation (see, e.g., [64]). Other methods, like the repetitive control concept (see, e.g., [33]) or the adaptive disturbance rejection (see, e.g., [65], [73]) do not distinguish between an identification and a compensation element.

Later, we will focus our attention on the latter strategy. The main goal of the eccentricity compensator is to eliminate the eccentricity induced disturbances in the strip exit thickness. But since the thickness gauge is located some distance behind the roll gap, the strip exit thickness signal h_{ex} has a significant time delay which depends on the strip exit velocity. This fact causes great problems in the development of a fast eccentricity compensation concept based on the strip exit thickness. Hence, we decided to use an indirect approach by suppressing the periodic vibrations in the piston force $F_h = A_1 p_1 - A_2 p_2$. If the deformation resistance of the strip material is sufficiently large, then this strategy also significantly decreases the periodic disturbances in the strip exit thickness. A very important fact for the commissioning engineer as well as for the maintenance staff is that the eccentricity compensator can be operated together with the conventional control concepts for HGC-systems and that it can be easily adjusted.

To design the eccentricity compensator, let us at first replace v by $v + \beta_T u$ with the supplementary plant input u in the non-linear HGC controller (5.15) and (5.16) of Proposition 5.1. Then by ignoring the dynamics of the mill stand, the closed-loop HGC-system written in deviations Δ around the stationary point (compare with (5.20)), now formulated in the state variables ΔF_h, Δx_k and Δv_k, reads as

$$\frac{d}{dt}\Delta F_h = -\alpha_1 f\left(\Delta x_k\right) - \frac{\partial}{\partial \Delta x_k} f\left(\Delta x_k\right)\Delta v_k + \beta_T u$$

$$\frac{d}{dt}\Delta x_k = \Delta v_k \tag{5.46}$$

$$\frac{d}{dt}\Delta v_k = \frac{1}{m_k}\left(\Delta F_h - d_k \Delta v_k + d\right)$$

with d as a periodic disturbance, with known period T summarizing the work and backup roll eccentricities, the moving mass $m_k = m_2 + m_3$ from Fig. 5.4 and the damping factor d_k. Since the deviations from the stationary point due to the roll eccentricities are very small, it is enough to consider the linearized system of (5.46) for the design of the eccentricity compensator. The system (5.46) linearized around $\Delta x^T = [\Delta F_h, \Delta x_k, \Delta v_k] = 0$ yields to

$$\frac{d}{dt}x = \begin{bmatrix} 0 & -\beta_T \alpha_1 \bar{c}_k & -\beta_T \bar{c}_k \\ 0 & 0 & 1 \\ \dfrac{1}{m_k} & 0 & \dfrac{-d_k}{m_k} \end{bmatrix} x + \begin{bmatrix} \beta_T \\ 0 \\ 0 \end{bmatrix} u + \begin{bmatrix} 0 \\ 0 \\ \dfrac{1}{m_k} \end{bmatrix} d \tag{5.47}$$

$$y = x_1$$

with

$$\bar{c}_k = \frac{A_1^2}{V_{0,1} + A_1 x_{k,ref}} + \frac{A_2^2}{V_{0,2} - A_2 x_{k,ref}} \tag{5.48}$$

and a lower bound for $\bar{c}_k$ given by (5.19). Now, condition (5.18) from Proposition 5.1 ensures that (5.47) is Hurwitz and even strictly positive real from u to y. These properties will subsequently help us to construct the eccentricity compensator.

In [73] a discrete version of the eccentricity compensator based on the projection theorem in a Hilbert space is presented. Here, we will use a passivity based argument for the time continuous case in order to prove the stability of the closed-loop. The main result for the adaptive cancellation of periodic disturbances with a known period but an unknown phase and amplitude is contained in the following proposition.

Proposition 5.2. *Consider a linear time-invariant system of the form*

$$\frac{d}{dt}x = Ax + bu + kd$$
$$y \quad = c^T x$$

(5.49)

where $x \in R^n$ is the state, $u \in R$ the input, $y \in R$ the measurable output and d a time-harmonic disturbance $d = D_{1,j}\sin\left(j\frac{2\pi}{T}t\right) + D_{2,j}\cos\left(j\frac{2\pi}{T}t\right)$, $j = 1,\dots,s$ with known period T and constant, but unknown $D_{1,j}, D_{2,j} \in R$. Let us assume that A is Hurwitz, (A, b) reachable, (A, c) observable and the corresponding transfer function $Z(s) = c^T (sI - A)^{-1} b$ is strictly positive real. Then the adaptive control law

$$u = \sum_{j=1}^{s}\left(U_{1,j}\sin\left(j\frac{2\pi}{T}t\right) + U_{2,j}\cos\left(j\frac{2\pi}{T}t\right)\right)$$

(5.50)

with

$$\frac{d}{dt}U_{1,j} = -\mu_j \sin\left(j\frac{2\pi}{T}t\right) y$$
$$\frac{d}{dt}U_{2,j} = -\mu_j \cos\left(j\frac{2\pi}{T}t\right) y$$

(5.51)

guarantees for $\mu_j > 0$, $j = 1,\dots,s$ that $y(t)$ exponentially decays to zero. The weights μ_j are used to adjust the convergence rate of the suppression of the various higher harmonics of the disturbance d.

Proof. Let $u^* = \sum_{j=1}^{s}\left(U_{1,j}^*\sin\left(j\frac{2\pi}{T}t\right) + U_{2,j}^*\cos\left(j\frac{2\pi}{T}t\right)\right)$, $U_{1,j}^*, U_{2,j}^* \in R$ denote the stationary input such that the stationary output $y^* = c^T x^* = 0$. Then the system (5.49) can be rewritten around the stationary solution (u^*, x^*) in the form

$$\frac{d}{dt}(x - x^*) = A(x - x^*) + b(u - u^*)$$
$$y - y^* \quad = c^T (x - x^*).$$

(5.52)

Since A is Hurwitz, (A, c) reachable, (A, c) observable and the transfer function $Z(s) = c^T (sI - A)^{-1} b$ is supposed to be strictly positive real, the Kalman-Yakubovich-Popov lemma (see Theorem 1.6 of Chapter 1 or e.g., [59], [144]) says that there exists a positive definite matrix P, a matrix L and an $\epsilon > 0$ such that the relations

$$PA + A^T P = -L^T L - \epsilon P$$
$$Pb \quad = c$$

(5.53)

hold. The closed-loop system (5.52) together with (5.50) and (5.51) in deviations from the stationary solution (u^*, x^*) reads as

$$\frac{d}{dt}\left(x - x^*\right) = A\left(x - x^*\right) + b\left\{\sum_{j=1}^{s}\left(U_{1,j} - U_{1,j}^*\right)\sin\left(j\tfrac{2\pi}{T}t\right) + \right.$$

$$\left. \sum_{j=1}^{s}\left(U_{2,j} - U_{2,j}^*\right)\cos\left(j\tfrac{2\pi}{T}t\right)\right\} \tag{5.54}$$

$$\frac{d}{dt}\left(U_{1,j} - U_{1,j}^*\right) = -\mu_j \sin\left(j\tfrac{2\pi}{T}t\right)c^T\left(x - x^*\right)$$

$$\frac{d}{dt}\left(U_{2,j} - U_{2,j}^*\right) = -\mu_j \cos\left(j\tfrac{2\pi}{T}t\right)c^T\left(x - x^*\right)$$

for $j = 1, \ldots, s$. Clearly, (5.54) is a linear time-varying system and in order to prove the exponential stability of the stationary solution $(U_{1,j}^*, U_{2,j}^*, x^*)$ with $y^* = c^T x^* = 0$, we make use of Theorem 1.3 of Chapter 1. We choose

$$V = \frac{1}{2}k_1\left(x - x^*\right)^T P\left(x - x^*\right) + \sum_{j=1}^{s}\frac{k_1}{2\mu_j}\sum_{i=1}^{2}\left(U_{i,j} - U_{i,j}^*\right)^2 \tag{5.55}$$

with $k_1 > 0$ and by calculating the time-derivative of V and taking into account the relations (5.53), we get

$$\frac{d}{dt}V = \frac{1}{2}k_1\left(x - x^*\right)^T\left(PA + A^T P\right)\left(x - x^*\right) + k_1\left(x - x^*\right)^T\left(Pb - c\right)$$

$$\left\{\sum_{j=1}^{s}\left(U_{1,j} - U_{1,j}^*\right)\sin\left(j\tfrac{2\pi}{T}t\right) + \sum_{j=1}^{s}\left(U_{2,j} - U_{2,j}^*\right)\cos\left(j\tfrac{2\pi}{T}t\right)\right\}.$$

$$= -\frac{1}{2}k_1\left(x - x^*\right)^T\left(L^T L + \epsilon P\right)\left(x - x^*\right) \leq 0. \tag{5.56}$$

Thus, from Theorem 1.3 of Chapter 1 it follows that $(U_{1,j}^*, U_{2,j}^*, x^*)$ is exponentially stable and hence y exponentially decays to zero. ∎

Remark 5.3. The result of the disturbance compensation concept of Proposition 5.2 remains valid even if the period T of the periodic disturbance changes to another stationary value. Furthermore, it turns out that if the variation of the period is sufficiently slow in comparison with the natural frequencies of the system, the disturbance compensation also shows good results in the transient case. Another important feature of the proposed concept is the fact that the disturbance controller (5.50) and (5.51) does not rely on the specific knowledge of the plant. It is enough for the plant to satisfy certain structural properties, namely stability and strict positive realness, which, in general, are not lost in the case of parameter variations.

Since (5.47) and (5.48) meet all the necessary requirements, Proposition 5.2 can be directly applied to design the eccentricity compensator. Fig. 5.9

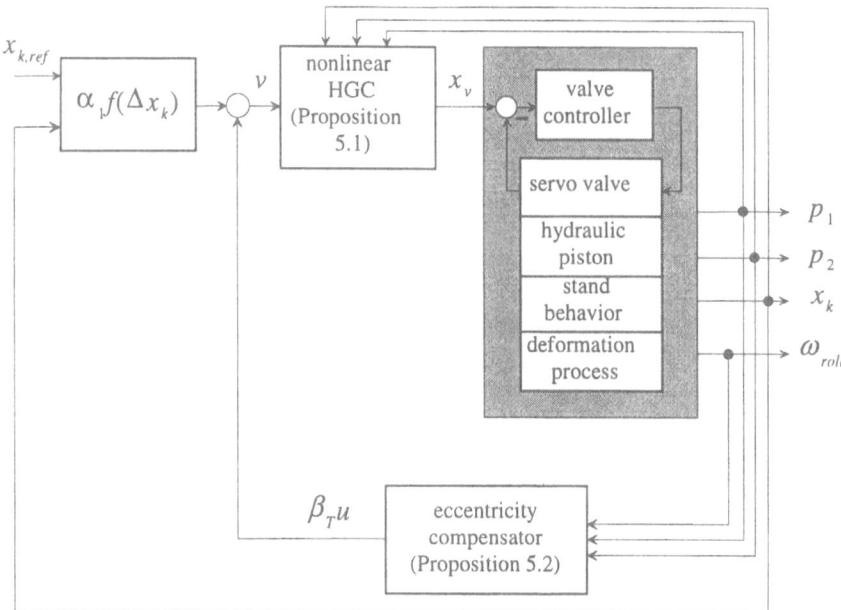

Fig. 5.9. Schematic diagram of the hydraulic position control concept in combination with the eccentricity compensator.

depicts the schematic diagram of the position control concept (HGC) from Proposition 5.1 with the plug-in eccentricity compensator of Proposition 5.2. Since the diameters of the backup and work rolls are known accurately it is possible to calculate the period of the backup and work roll eccentricities by measuring the angular velocity ω_{roll} of at least one of the rolls, usually a work roll. Of course, this takes for granted that there is no slipping between the work and backup rolls. To test the control concept with the eccentricity compensation, the same mill simulator as described in Subsection 5.2.4 is used. The reference input for the position control is chosen as $x_{k,ref} = 100 \cdot 10^{-6} \sigma (t - 1)$ and in addition to Subsection 5.2.4 the backup and the work roll eccentricities of (5.33) are set to $x_{e,B}(t) = 30 \cdot 10^{-6} \sin(17.1t + \pi/4)$ and $x_{e,W}(t) = 25 \cdot 10^{-6} \sin(31.75t + \pi/2)$. One can see from Fig. 5.10 that the eccentricity induced periodic disturbances in the deviation of the strip exit thickness Δh_{ex} (see (5.38)) are significantly reduced by the eccentricity compensator. Various other simulations and field tests [73] prove that the strategy of eliminating the periodic disturbances in the hydraulic force really does lead to an improvement in the strip exit thickness deviation. The simulation results of the spool valve displacement x_v and the deviation of the displacement of the hydraulic piston Δx_k (see (5.38)) in Fig. 5.10 should demonstrate the effect of the eccentricity compensator on the hydraulic adjustment system. Note that in this connection a Δ before a quantity always refers to the deviation of this quantity from the nominal operating point.

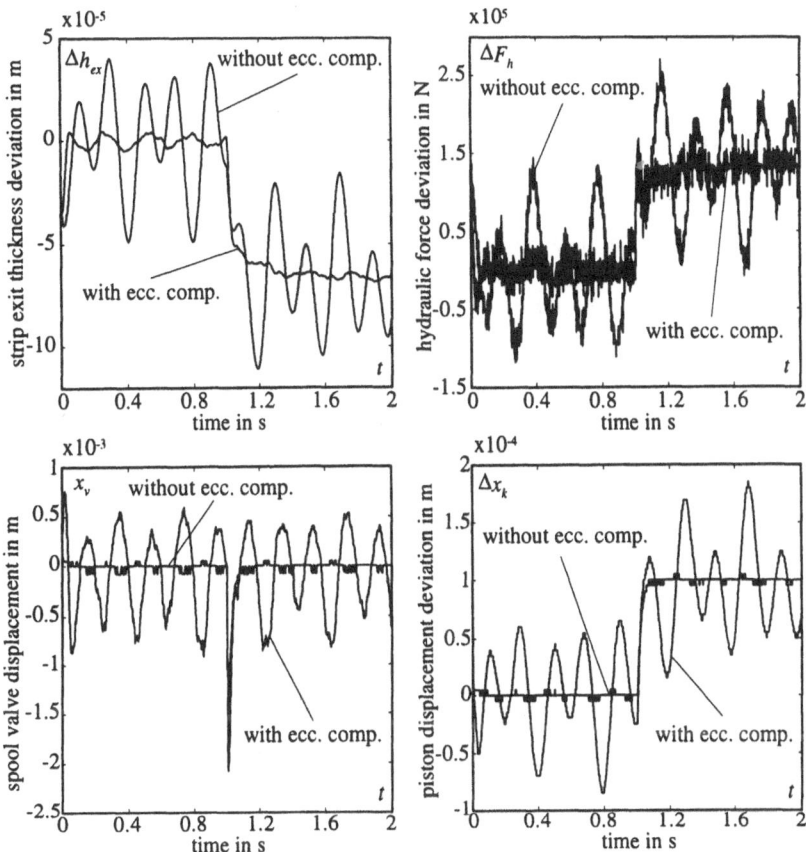

Fig. 5.10. Simulation of the exit thickness, the hydraulic force, the spool valve position and the piston displacement with and without eccentricity compensation.

Remark 5.4. There are many different techniques presented in the literature for the active suppression and rejection of periodic disturbances. Essentially, these approaches differ from the assumptions on the structural information of the system and on the stochastic information of the disturbance. Nevertheless, there are, of course, essential similarities between the different algorithms, as is shown in [134]. An adaptive digital plug-in algorithm for rejecting periodic disturbances with known frequency and its application to a magnet disc drive experiment is considered in [46]. Here, the reference signal is assumed to be periodic with the same period as the disturbance. The LMS (Least Mean

Squares) adaptive algorithm used is driven by an error signal coming from a moving average filter. Another interesting approach for adaptive disturbance rejection, extended by the factorization approach over the set of BIBO-stable transfer functions, is presented in [43] and it is also worth consulting the references cited there.

5.4 Pump-displacement-controlled Rotational Piston Actuator

Fig. 5.11 illustrates the cross section of a hydrostatic unit with a variable-displacement axial-piston pump and a fixed-displacement axial-piston motor. Typical applications for this type of hydrostatic drives are ship steering sys-

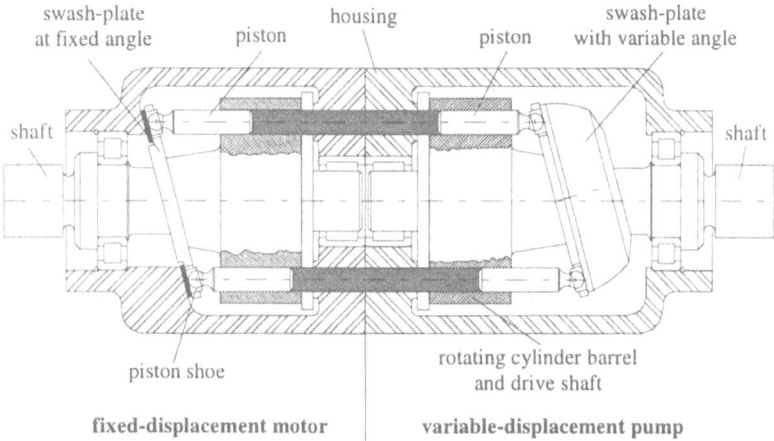

Fig. 5.11. Cross section of a hydrostatic unit consisting of a variable-displacement pump and a fixed-displacement motor.

tems, antenna drives or vehicle drive systems (see, e.g., [78]). In these systems, the pump converts mechanical energy into fluid energy and the motor converts the fluid energy back into mechanical energy. Since the swash-plate angle of the pump is adjustable both in the positive as well as in the negative direction the hydraulic transmission offers continuously variable output speed in all four quadrants. Therefore, the differentiation between motor and pump is arbitrary and does not necessarily match with the real operating conditions. The transmission lines which connect the pump and the motor are assumed to be short and we assume that the low- and high-pressure side (transmission line + chambers in the pump or motor + connecting passages) can be represented by one pressure value each. Also here, the operating conditions determine which side has high and which side has low pressure. There are two main sources of losses in the hydrostatic drive unit, namely losses due to internal and external leakage flows and losses due to friction. The friction losses cause the oil to heat up and so the oil temperature rises. To counteract this effect a flush valve drains oil from the low pressure side. A so-called replenishing system consisting of a replenishing pump and two check valves supplies the low-pressure side with cooled, filtered oil and makes up for the leakage flows and the flush flow. The check valves prevent the pressure on both sides from falling below the replenishing pressure p_R. Furthermore, safety relief valves are used to restrict the pressure values in both sides to a predefined maximum.

5.4.1 Mathematical Model of the Pump-motor-unit

For the derivation of the mathematical model of the hydrostatic drive unit, some simplifying assumptions will be made, which constitute no essential restriction for the controller design:

- In view of the application to be considered we assume that the pump-motor configuration is close coupled, i.e. the transmission lines between the pump and the motor are rather short. Therefore, we may ignore the dynamics caused by the transport of the oil via the transmission lines from the pump to the motor and *vice versa*.
- The pressures, p_1 and p_2, of the two pressure sides are supposed not to exceed the maximum value defined by the safety relief valves.
- It is assumed that the replenishing pump perfectly compensates for the drained oil from the flush valve. At this point it is worth mentioning that in reality the flush valve and the check valves of the replenishing system have a non-linear opening characteristics and a non-negligible dynamics. However, if the dynamics are sufficiently fast they may be ignored in the controller design. Of course, in the simulation model for testing the controllers all these effects should be included.

In order to get a better understanding, the required quantities for the mathematical model of the motor are shown in Fig. 5.12. Henceforth, an index p or m always refers to the corresponding quantity of the pump or motor, respectively. Since hydrostatic pumps (motors) are designed by a finite number N_p (N_m) of pistons, the flow from the pump to the transmission line (from the transmission line to the motor) depends on the angle of rotation. These fluctuations in the flow also cause torque ripples, which have, e.g. in the case $N_p = 9$, an amplitude of approximately 1.5 % of the normalized torque [90]. For further details of the range of constructions and the corresponding acting forces and torques of hydrostatic pumps and motors, the reader is also referred to [54]. Subsequently, we will use an average pump and motor flow q_p and q_m of the form

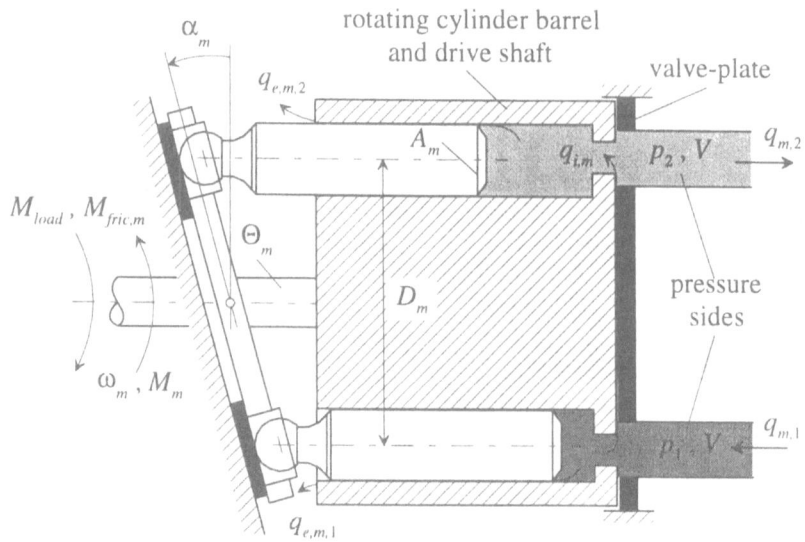

Fig. 5.12. Schematic diagram of the fixed-displacement axial-piston motor.

$$q_p = \frac{N_p}{2\pi} A_p D_p \tan(\alpha_p) \omega_p \quad \text{and} \quad q_m = \frac{N_m}{2\pi} A_m D_m \tan(\alpha_m) \omega_m \quad (5.57)$$

with the angular velocity ω, the maximum geometric displacement D, the effective piston area A and the stroke angle α. The generated pump and motor torque M_p and M_m are calculated by considering an ideal pump or motor, which means a lossless energy conversion. Thus, we obtain an average torque by

$$M_j = \frac{N_j}{2\pi} A_j D_j \tan(\alpha_j)(p_1 - p_2) \quad \text{with} \quad j \in \{p, m\}. \quad (5.58)$$

From Fig. 5.12 we see that the average motor flow q_m due to (5.57) is associated with the flow from the transmission line to the motor, $q_{m,1}$, and the flow from the motor to the transmission line, $q_{m,2}$, in the form

$$
\begin{aligned}
q_{m,1} &= q_m + q_{i,m} + q_{e,m,1} \\
q_{m,2} &= q_m + q_{i,m} - q_{e,m,2},
\end{aligned}
\qquad (5.59)
$$

with the internal and external leakage flows of the motor $q_{i,m}$, $q_{e,m,1}$ and $q_{e,m,2}$. Analogous to (5.58) the average pump flow q_p is given by

$$
\begin{aligned}
q_{p,1} &= q_p - q_{i,p} - q_{e,p,1} \\
q_{p,2} &= q_p - q_{i,p} + q_{e,p,2},
\end{aligned}
\qquad (5.60)
$$

with $q_{p,1}$ as the flow from the pump to the transmission line, $q_{p,2}$ as the flow from the transmission line to the pump and the internal and external leakage flows of the pump $q_{i,p}$, $q_{e,p,1}$ and $q_{e,p,2}$. In contrast to the valve-controlled translational piston actuator of Section 5.1, here the heating of the oil and hence the dependence on the temperature T will be taken into account. Thus, by assuming rigid connection lines, we get the continuity equations for the two pressure sides in the form

$$\frac{\mathrm{d}}{\mathrm{d}t}\left(\rho_{oil}\left(p_1,T\right)V\right) = \rho_{oil}\left(p_1,T\right)\left(q_{p,1}-q_{m,1}\right)$$
$$\frac{\mathrm{d}}{\mathrm{d}t}\left(\rho_{oil}\left(p_2,T\right)V\right) = \rho_{oil}\left(p_2,T\right)\left(q_{m,2}-q_{p,2}\right), \tag{5.61}$$

where V denotes the total volume (transmission line + chambers + connecting passages) of each pressure side and $q_{p,1}$, $q_{m,1}$, $q_{p,2}$, $q_{m,2}$ are given by (5.59) and (5.60), see, e.g., [13], [96]. The equations of motion for the pump and the motor read as

$$\Theta_p\frac{\mathrm{d}}{\mathrm{d}t}\omega_p = M_{drive} - M_p - M_{fric,p}$$
$$\Theta_m\frac{\mathrm{d}}{\mathrm{d}t}\omega_m = M_m - M_{load} - M_{fric,m}, \tag{5.62}$$

with the angular velocity ω, the friction torque M_{fric}, the drive torque of the pump M_{drive}, the load torque of the motor M_{load} and the moment of inertia Θ of the pump or motor plus all parts rigidly connected to the pump or motor shaft, respectively [13], [96].

The leakage flows and the friction torques have an essential influence on the efficiency of the hydrostatic drive unit. One can find many models in the literature. For a comparison of three different models, see, e.g., [138]. Due to the geometry of the leakage paths, the internal and external leakage flows of (5.59) and (5.60) have basically the same laminar characteristic and hence are linear functions of the pressure difference or pressure

$$q_{i,m} = \frac{C_{i,m}}{\mu\left(T\right)}\left(p_1-p_2\right), \quad q_{e,m,1} = \frac{C_{e,m}}{\mu\left(T\right)}p_1, \quad q_{e,m,2} = \frac{C_{e,m}}{\mu\left(T\right)}p_2$$
$$q_{i,p} = \frac{C_{i,p}}{\mu\left(T\right)}\left(p_1-p_2\right), \quad q_{e,p,1} = \frac{C_{e,p}}{\mu\left(T\right)}p_1, \quad q_{e,p,2} = \frac{C_{e,p}}{\mu\left(T\right)}p_2, \tag{5.63}$$

with the positive leakage coefficients $C_{i,m}$, $C_{i,p}$, $C_{e,m}$, $C_{e,p}$ and the dynamic viscosity $\mu\left(T\right)$. The dynamic viscosity $\mu\left(T\right)$ changes markedly with the temperature T and can be fairly closely approximated by

$$\mu\left(T\right) = \mu_0\exp\left(-\lambda_1\left(T-T_0\right)\right), \tag{5.64}$$

where μ_0 is the viscosity at the reference temperature T_0 and λ_1 is a characteristic constant of the type of oil used [13], [96]. In general, the friction torque M_{fric} is assumed to consist of two different parts, namely

$$M_{fric,j} = k_{d,j}\mu(T)\,\omega_j + k_{p,j}(p_1 + p_2)\,\text{sign}(\omega_j) \quad , \quad j \in \{p, m\} \qquad (5.65)$$

with the positive friction coefficients $k_{d,j}$ and $k_{p,j}$ depending on the geometry and the specific construction. The first part, which is proportional to the angular velocity ω and the dynamic viscosity $\mu(T)$, is mainly caused by the shearing of the fluid in the small clearances and the second part is a consequence of the forces acting on the swash-plate [54], [146]. In some literature the friction coefficient k_p is not supposed to be constant, as in (5.65), but is assumed to be proportional to the stroke angle α of the swash-plate (e.g., [96]). However, no friction model optimally fits reality and the parameters have to be adjusted to the measurements in each individual case. For the purpose of a controller design, henceforth, we will neglect the second part of the friction torque, i.e. $k_{p,p} = k_{p,m} = 0$. Of course, in order to get realistic simulation results, the whole friction model of (5.65) extended by the well known stiction phenomenon should be included in the simulation model.

Summarizing (5.57) - (5.65) and taking into account that the change of temperature T is negligible in comparison to all other dynamic parts of the system, i.e. $\frac{d}{dt}T = 0$, we are able to formulate the mathematical model of the hydrostatic drive unit of Fig. 5.11 in the form

$$\frac{V_1}{\beta_T(T)}\frac{d}{dt}p_1 = \frac{N_p}{2\pi}A_pD_p\tan(\alpha_p)\,\omega_p - \frac{N_m}{2\pi}A_mD_m\tan(\alpha_m)\,\omega_m -$$
$$\frac{C_i}{\mu(T)}(p_1 - p_2) - \frac{C_e}{\mu(T)}p_1$$
$$\frac{V_2}{\beta_T(T)}\frac{d}{dt}p_2 = \frac{N_m}{2\pi}A_mD_m\tan(\alpha_m)\,\omega_m - \frac{N_p}{2\pi}A_pD_p\tan(\alpha_p)\,\omega_p +$$
$$\frac{C_i}{\mu(T)}(p_1 - p_2) - \frac{C_e}{\mu(T)}p_2$$
$$\Theta_p\frac{d}{dt}\omega_p = M_{drive} - \frac{N_p}{2\pi}A_pD_p\tan(\alpha_p)(p_1 - p_2) - k_{d,p}\mu(T)\,\omega_p$$
$$\Theta_m\frac{d}{dt}\omega_m = \frac{N_m}{2\pi}A_mD_m\tan(\alpha_m)(p_1 - p_2) - M_{load} - k_{d,m}\mu(T)\,\omega_m$$

$$(5.66)$$

where $C_e = C_{e,p} + C_{e,m}$, $C_i = C_{i,p} + C_{i,m}$ and $\beta_T(T)$ is the bulk modulus of oil (see (5.2)), with the temperature T as a parameter. In the temperature range which is practically relevant the compressibility $\beta_T(T)$ as a function of the temperature T can be approximated by

$$\beta_T(T) = \beta_{T,0} + \lambda_2(T - T_0), \qquad (5.67)$$

where $\beta_{T,0}$ is the bulk modulus at the reference temperature T_0 and λ_2 is a characteristic constant of the type of oil used [96], [101]. All other quantities of (5.66) have been explained in the equations above.

As already mentioned before, the replenishing system prevents the pressure on both sides, p_1 and p_2, from falling below the replenishing pressure

p_R. Therefore, we have to distinguish between three different operating situations:

Case I: The pressure p_2 is held at the constant value of p_R, i.e. $\frac{d}{dt}p_2 = 0$, and the load changes are performed only by the pressure p_1. Since in general p_R is very small, we may set p_R to zero and hence we have $\Delta p = p_1 - p_2 \approx p_1$. Thus, the continuity equation for the pressure side p_1 written in the pressure difference Δp read as

$$\frac{V}{\beta_T(T)}\frac{d}{dt}\Delta p = \frac{N_p}{2\pi}A_pD_p\tan(\alpha_p)\omega_p - \frac{N_m}{2\pi}A_mD_m\tan(\alpha_m)\omega_m$$
$$-\frac{C_i+C_e}{\mu(T)}\Delta p. \tag{5.68}$$

Case II: Analogous to the first case, the pressure $p_1 = p_R \approx 0$ and only the pressure p_2 is varying. Then the continuity equation for the pressure side p_2 written in the pressure difference $\Delta p = p_1 - p_2 \approx -p_2$ yields to

$$\frac{V}{\beta_T(T)}\frac{d}{dt}\Delta p = \frac{N_p}{2\pi}A_pD_p\tan(\alpha_p)\omega_p - \frac{N_m}{2\pi}A_mD_m\tan(\alpha_m)\omega_m$$
$$-\frac{C_i+C_e}{\mu(T)}\Delta p. \tag{5.69}$$

Case III: The third case covers the situation where both pressures, p_1 and p_2, are varying simultaneously. This can happen only, if there is a change from case I to case II or *vice versa*. By subtracting the continuity equations for the two pressure sides, we get

$$\frac{V}{2\beta_T(T)}\frac{d}{dt}\Delta p = \frac{N_p}{2\pi}A_pD_p\tan(\alpha_p)\omega_p - \frac{N_m}{2\pi}A_mD_m\tan(\alpha_m)\omega_m$$
$$-\frac{C_i+C_e/2}{\mu(T)}\Delta p \tag{5.70}$$

with $\Delta p = p_1 - p_2$.

Thus, from (5.68) - (5.70) together with (5.66) it can be easily seen that the mathematical model of the hydrostatic drive unit, which covers all three operating situations, can be written as a PCHD-system (see Chapter (1))

$$\frac{d}{dt}x = (J(x) + J_u(x)u - S(x))\left(\frac{\partial V}{\partial x}\right)^T + G_d(x)d \tag{5.71}$$

with the state $x^T = \left[\dfrac{\tilde{V}}{\beta_T(T)}\Delta p, \Theta_p\omega_p, \Theta_m\omega_m\right]$, the exogenous input $d^T = [M_{load}, M_{drive}]$, the plant input $u = \tan(\alpha_p)$, the positive definite storage function $V(x)$

$$V(x) = \frac{1}{2}\left(\frac{\tilde{V}}{\beta_T(T)}\Delta p^2 + \Theta_p\omega_p^2 + \Theta_m\omega_m^2\right), \tag{5.72}$$

the positive definite, symmetric matrix

$$S\left(x\right) = \begin{bmatrix} \dfrac{\tilde{C}}{\mu\left(T\right)} & 0 & 0 \\ 0 & k_{d,p}\mu\left(T\right) & 0 \\ 0 & 0 & k_{d,m}\mu\left(T\right) \end{bmatrix} \tag{5.73}$$

the skew symmetric matrices

$$J\left(x\right) = \frac{N_m}{2\pi} A_m D_m \tan\left(\alpha_m\right) \begin{bmatrix} 0 & 0 & -1 \\ 0 & 0 & 0 \\ 1 & 0 & 0 \end{bmatrix}, \; J_u\left(x\right) = \frac{N_p}{2\pi} A_p D_p \begin{bmatrix} 0 & 1 & 0 \\ -1 & 0 & 0 \\ 0 & 0 & 0 \end{bmatrix} \tag{5.74}$$

and the matrix for the exogenous inputs

$$G_d^T\left(x\right) = \begin{bmatrix} 0 & 0 & -1 \\ 0 & 1 & 0 \end{bmatrix}. \tag{5.75}$$

For the cases I and II, (5.68) and (5.69), $\tilde{V} = V$ and $\tilde{C} = C_i + C_e$, and for case III (5.70) $\tilde{V} = \frac{V}{2}$ and $\tilde{C} = C_i + \frac{C_e}{2}$.

Remark 5.5. The model of the hydrostatic drive unit of Fig. 5.11 has the same mathematical structure as the Ćuk-converter of Subsection 3.4.2. Here also, the control input $\tan\left(\alpha_p\right)$, with α_p as the stroke angle of the pump, influences only the internal energy flow and thus does not change the total amount of energy stored in the system.

The next table, Table 5.2, summarizes the nomenclature used for the pump-motor-unit, where an index p or m always refers to the corresponding quantity of the pump or motor, respectively.

In vehicular drive systems a typical control application for the hydrostatic drive unit is to track a desired trajectory of the angular velocity of the motor ω_m in the presence of load and/or drive torque variations. Beside the classical control approaches (see, e.g., [78] and the literature cited there), it is obvious that we may fall back on all the results known from literature for the dc-to-dc converters with the same mathematical structure. For this purpose, the reader is referred to Sections 2.2, 3.5 and the literature cited in it, especially [26], [107], [132]. In the following, we do not go into the details of the controller design for hydrostatic drive units more closely, but will focus on another problem. Usually, the variable stroke angle α_p of the pump-swashplate is considered as the plant input, as it is in (5.71), and the dynamics of the swash-plate and the corresponding control unit are neglected. However, this simplification is admissible if the dynamics of the swash-plate are fast in relation to the pump-motor dynamics. On condition that α_p is measurable, this can possibly be achieved by means of a fast acting inner control

Table 5.2. Nomenclature for the pump-motor-unit.

A_m, A_p	:	effective piston areas
$C_{e,p}, C_{e,m}, C_e$	:	external laminar leakage coefficients
$C_{i,p}, C_{i,m}, C_i$	:	internal laminar leakage coefficients
$\tilde{C}$	:	laminar leakage coefficient
D_m, D_p	:	maximum geometric displacement
$k_{d,m}, k_{d,p}$	:	viscous friction coefficients
$k_{p,m}, k_{p,p}$	:	friction coefficients
M_{drive}	:	drive torque of pump
$M_{fric,p}, M_{fric,m}$	:	friction torques
M_{load}	:	load torque of motor
M_m, M_p	:	average torques of pump/motor
N_m, N_p	:	number of pistons
p_1, p_2	:	pressures of the two pressure sides
q_m, q_p	:	average flows of pump/motor
$q_{e,p,1}, q_{e,p,2}, q_{e,m,1}, q_{e,m,2}$	:	external leakage flows
$q_{i,p}, q_{i,m}$	:	internal leakage flows
T	:	oil temperature
T_0	:	reference temperature
$V, \tilde{V}$	:	volumes of the two pressure sides
α_m, α_p	:	stroke angle of swash-plates
$\beta_T(T)$	:	bulk modulus of oil at temperature T
$\beta_{T,0}$	:	bulk modulus of oil at temperature T_0
Θ_m, Θ_p	:	sum of rigidly connected moments of inertia
λ_1, λ_2	:	characteristic oil coefficients
$\mu(T)$	:	dynamic viscosity of oil at temperature T
μ_0	:	dynamic viscosity of oil at temperature T_0
ρ_{oil}	:	density of oil
ω_m, ω_p	:	angular velocities

loop for α_p. Otherwise the mathematical model of the swash-plate has to be taken into account. For reasons of reliability, in many industrial applications no measurement device is planned for the swash-plate angle. Therefore, the information about the swash-plate angle has to be obtained otherwise, e.g. by means of an observer. This topic will be treated in more detail in the following section.

5.5 Discrete Open-loop Observer for the Swash-plate Angle

Fig. 5.13 depicts the schematic diagram of the swash-plate mechanism of a variable-displacement pump. By means of the two piston forces $F_A = A_A p_A$ and $F_B = A_B p_B$ of the two actuators A and B and the spring force F_S the swash-plate angle α_p can be controlled in a range $-\alpha_{p,\max} \leq \alpha_p \leq \alpha_{p,\max}$. A hydro-mechanical feedback mechanism opens the valve of the corresponding actuator with an orifice area $A_o\,(\Delta\alpha_p)$ and connects the chamber either with the tank at pressure p_T or with the supply pressure p_S. This is done in such a way that the error $\Delta\alpha_p = \alpha_p - \alpha_{p,d}$ between the actual and the desired swash-plate angle α_p and $\alpha_{p,d}$ is eliminated.

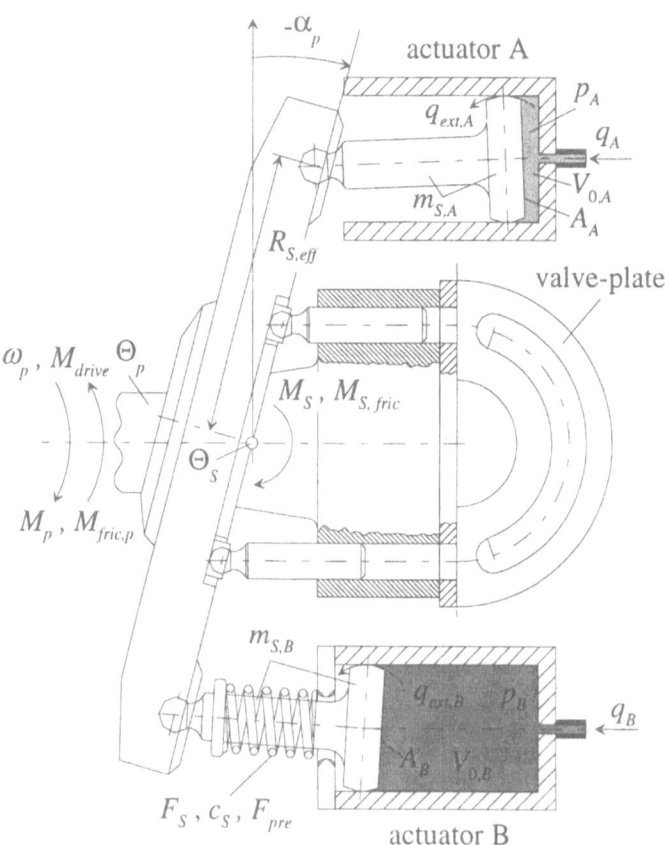

Fig. 5.13. Schematic diagram of the swash-plate mechanism of the variable-displacement pump.

5.5.1 Mathematical Model of a Swash-plate Mechanism

An excellent description of the closed-form equations of a variable-displacement pump can be found in [87] and [89]. Here, the attention is mainly directed to the construction of the swash-plate mechanism and the influence of parameter variations on its dynamic behavior. The objective of this subsection is the derivation of a detailed mathematical model that captures all the essential dynamic properties of the system and that serves as a basis for the observer design for the swash-plate angle α_p [71].

Mechanical Part. Calculating the rate of change of the angular momentum around the swash-plate pivot, we get

$$
\begin{aligned}
\frac{d}{dt}\alpha_p &= v_p \\
\Theta_S \frac{d}{dt} v_p &= (F_{S,A} - F_{S,B} - F_S)\, R_{S,eff} \cos(\alpha_p) - M_{S,fric} + M_S
\end{aligned}
\tag{5.76}
$$

with the moment of inertia of the swash-plate Θ_S, the effective distance of the acting forces and the swash-plate pivot $R_{S,eff}$, the forces $F_{S,A}$ and $F_{S,B}$ from the actuator A and B given by

$$
\begin{aligned}
F_{S,A} &= -m_{S,A}\frac{d^2}{dt^2}\left(R_{S,eff}\sin(\alpha_p)\right) + F_A \\
F_{S,B} &= m_{S,B}\frac{d^2}{dt^2}\left(R_{S,eff}\sin(\alpha_p)\right) + F_B,
\end{aligned}
\tag{5.77}
$$

the spring force

$$
F_S = F_{pre} + c_S R_{S,eff}\left(\sin(\alpha_p) + \sin(\alpha_{p,max})\right),
\tag{5.78}
$$

the friction torque $M_{S,fric}$ and the so-called swivel torque M_S, which is naturally induced by the pump [87], [89]. In (5.77) $m_{S,A}$ and $m_{S,B}$ denote the sum of the masses of the piston and the piston rod of the two actuators A and B, respectively, and in (5.78) F_{pre} stands for the prestress-force of the spring and c_S is the spring coefficient. Thus, the equations of motion can be summarized in the form

A.) | end stop $\alpha_{p,max}$: |

If $F_A - F_B > F_{pre} + 2c_S R_{S,eff}\sin(\alpha_{p,max})$ then

$$\alpha_p = \alpha_{p,max} \quad \text{and} \quad v_p = 0$$

B.) | end stop $-\alpha_{p,max}$: |

If $F_A - F_B < F_{pre}$ then

$$\alpha_p = -\alpha_{p,max} \quad \text{and} \quad v_p = 0$$

C.) operating region $-\alpha_{p,\max} \le \alpha_p \le \alpha_{p,\max}$:

If $-\alpha_{p,\max} \le \alpha_p \le \alpha_{p,\max}$ then

$$\frac{d}{dt}\alpha_p = \nu_p$$

$$\frac{d}{dt}\nu_p = \frac{R_{S,eff}\cos(\alpha_p)(F_A - F_B - F_S) - M_{S,fric} + M_S}{\left(\Theta_S + (m_{S,A} + m_{S,B})R_{S,eff}^2\cos^2(\alpha_p)\right)} +$$

$$\frac{(m_{S,A} + m_{S,B})R_{S,eff}^2\sin(\alpha_p)\cos(\alpha_p)\nu_p^2}{\left(\Theta_S + (m_{S,A} + m_{S,B})R_{S,eff}^2\cos^2(\alpha_p)\right)} \qquad (5.79)$$

with F_S from (5.78).

Hydraulic Part. The actuators A and B of Fig. 5.13 are single-ended, single-acting hydraulic rams. Under the assumption that the change of the temperature T is slow in comparison to the dynamics of the hydro-mechanical part, i.e. $\frac{d}{dt}T = 0$, the continuity equations for the two chambers read as

$$\frac{(V_{0,A} + A_A R_{S,eff}\sin(\alpha_p))}{\beta_T(T)}\frac{d}{dt}p_A = q_A - A_A R_{S,eff}\cos(\alpha_p)\nu_p - q_{ext,A}$$

$$\frac{(V_{0,B} - A_B R_{S,eff}\sin(\alpha_p))}{\beta_T(T)}\frac{d}{dt}p_B = q_B + A_B R_{S,eff}\cos(\alpha_p)\nu_p - q_{ext,B}$$

$$(5.80)$$

with the pressures p_A and p_B in the two chambers, the effective piston areas A_A and A_B, the volumes $V_{0,A}$ and $V_{0,B}$ for $\alpha_p = 0$, the bulk modulus $\beta_T(T)$ with the temperature T as a parameter from (5.67) and the laminar external leakage flows

$$q_{ext,A} = \frac{C_{ext,A}}{\mu(T)}p_A \quad \text{and} \quad q_{ext,B} = \frac{C_{ext,B}}{\mu(T)}p_B. \qquad (5.81)$$

Here, $C_{ext,A}$ and $C_{ext,B}$ denote the laminar leakage coefficients and $\mu(T)$ is the dynamic viscosity as a function of the temperature T due to (5.64). The flows from and to the valves of the two chambers, q_A and q_B, are determined by a hydro-mechanical feedback mechanism in such a way that the error $\Delta\alpha_p = \alpha_p - \alpha_{p,d}$ between the actual and the desired swash-plate angle α_p and $\alpha_{p,d}$ becomes zero. This is achieved by connecting the corresponding chamber with an orifice area $A_o(\Delta\alpha_p)$ with the tank or supply pressure. Thus, the flows are given by

A.) $\Delta\alpha_p > 0$:

$$q_A = C_d\sqrt{\frac{2}{\rho_{oil}(T)}}A_o(\Delta\alpha_p)\sqrt{p_S - p_A}$$

$$q_B = -C_d\sqrt{\frac{2}{\rho_{oil}(T)}}A_o(\Delta\alpha_p)\sqrt{p_B - p_T}$$

$$(5.82)$$

B.) $\boxed{\Delta\alpha_p < 0:}$

$$q_A = -C_d\sqrt{\frac{2}{\rho_{oil}(T)}}A_o(\Delta\alpha_p)\sqrt{p_A - p_T}$$
$$q_B = C_d\sqrt{\frac{2}{\rho_{oil}(T)}}A_o(\Delta\alpha_p)\sqrt{p_S - p_B},$$

(5.83)

where p_T and p_S denote the tank and supply pressure, respectively, C_d is the discharge coefficient and $\rho_{oil}(T)$ is the density of oil, now with the temperature T as a parameter. Similar to the bulk modulus $\beta_T(T)$, the density of oil $\rho_{oil}(T)$ can be fairly closely approximated by an affine relationship

$$\rho_{oil}(T) = \rho_{oil,0} + \lambda_3(T - T_0),$$

(5.84)

where $\rho_{oil,0}$ is the density at the reference temperature T_0 and λ_3 is a characteristic constant of the type of oil used [96], [101]. For slit-type sharp-edged orifices the discharge coefficient may be set $C_d \approx 0.6$, regardless of the particular geometry [96], [101].

5.5.2 Model Simplification Based on Physical Considerations

Now, for certain types of applications it is possible to simplify the complexity of the mathematical model of the swash-plate mechanism in such a way that the resulting model still contains all the essential dynamic effects. In a first step, let us assume that the two actuators A and B have the identical geometry, i.e. $A_A = A_B = A$ and $V_{0,A} = V_{0,B} = V_0$. Further, if we neglect the external leakage flows $q_{ext,A} = q_{ext,B} = 0$, then the flows q_A and q_B from (5.82) and (5.83) follow the relation $q_A = -q_B$. Thus, with the assumption $p_T = 0$, we get $p_A + p_B = p_S$ and we are able to formulate (5.82) and (5.83) as functions of the load pressure $\Delta p = p_A - p_B$ by

A.) $\boxed{\Delta\alpha_p > 0:}$

$$q_A = -q_B = C_d\sqrt{\frac{2}{\rho_{oil}(T)}}A_o(\Delta\alpha_p)\sqrt{\frac{p_S - \Delta p}{2}}$$

(5.85)

B.) $\boxed{\Delta\alpha_p < 0:}$

$$q_A = -q_B = -C_d\sqrt{\frac{2}{\rho_{oil}(T)}}A_o(\Delta\alpha_p)\sqrt{\frac{p_S + \Delta p}{2}}$$

(5.86)

Additional investigations have shown that a Taylor series approximation up to the first order of (5.85) and (5.86) around $\Delta p = 0$ suffices for the description of the dynamics of the system. Thus we get

A.) $\boxed{\Delta\alpha_p > 0 :}$

$$q_A = -q_B = \frac{C_d}{\sqrt{\rho_{oil}(T)}} A_o(\Delta\alpha_p) \left(\sqrt{p_S} - \frac{1}{2\sqrt{p_S}}\Delta p\right) \quad (5.87)$$

B.) $\boxed{\Delta\alpha_p < 0 :}$

$$q_A = -q_B = -\frac{C_d}{\sqrt{\rho_{oil}(T)}} A_o(\Delta\alpha_p) \left(\sqrt{p_S} + \frac{1}{2\sqrt{p_S}}\Delta p\right) . \quad (5.88)$$

Since in (5.80) the expressions on the left hand side are very small due to $1/\beta_T(T)$, we can replace the dynamic equations (5.80) by their quasi-steady-state representation in the sense of the singular perturbation theory (see, e.g., [59]). Thus, the differential equations (5.80) degenerate into equations and by inserting (5.87) and (5.88) into these equations, we obtain an expression for the load pressure Δp of the form

$$\Delta p = -\frac{2\sqrt{\rho_{oil}(T)\, p_S} A R_{S,eff}}{C_d A_o(\Delta\alpha_p)} \cos(\alpha_p)\nu_p + 2p_S \operatorname{sign}(\Delta\alpha_p) . \quad (5.89)$$

In a second step, our analysis shows that in (5.79) the masses $m_{S,A}$ and $m_{S,B}$ of the piston and the piston rod of the two actuators A and B, as well as the friction and the swivel torque $M_{S,fric}$ and M_S, make only a minor contribution to the torque balance. Furthermore, in the considered operating range of the swash-plate angle, in our case $-21.5\frac{\pi}{180} \leq \alpha_p \leq 21.5\frac{\pi}{180}$, all expressions in α_p may be linearized around $\alpha_p = 0$, i.e. $\cos(\alpha_p) \cong 1$, $\sin(\alpha_p) \cong \alpha_p$ and $\sin(\alpha_{p,max}) \cong \alpha_{p,max}$. With these simplifications (5.79) together with (5.78) becomes

$$\frac{d}{dt}\alpha_p = \nu_p$$
$$\Theta_S\frac{d}{dt}\nu_p = R_{S,eff}(A\Delta p - F_{pre} - c_S R_{S,eff}(\alpha_p + \alpha_{p,max})) . \quad (5.90)$$

Using the same procedure as for the hydraulic part, we will regard the moment of inertia of the swash-plate Θ_S in (5.90) as a perturbation parameter. In [89] and in the literature cited there it is also considered that the inertia of the swash-plate is negligible to the stiffness of the servo-system. Substituting (5.89) in the quasi-steady-state representation of (5.90), we end up with

$$\frac{d}{dt}\alpha_p = \frac{C_d A_o(\Delta\alpha_p)}{2\sqrt{\rho_{oil}(T)\, p_S} A^2 R_{S,eff}} (-c_S R_{S,eff}\alpha_p + 2Ap_S \operatorname{sign}(\Delta\alpha_p)$$
$$- F_{pre} - c_S R_{S,eff}\alpha_{p,max}) . \quad (5.91)$$

Due to the fact that in general in (5.91) the expression $2Ap_S$ is much bigger than $-F_{pre} - c_S R_{S,eff} \alpha_{p,\max}$, another simplification, but now the last one, is possible

$$\frac{d}{dt}\alpha_p = \underbrace{-\frac{C_d A_o\left(\Delta\alpha_p\right) c_S}{2A^2 \sqrt{\rho_{oil}\left(T\right) p_S}}}_{\zeta_1}\alpha_p + \underbrace{\frac{C_d A_o\left(\Delta\alpha_p\right)}{A R_{S,eff}}\sqrt{\frac{p_S}{\rho_{oil}\left(T\right)}}}_{\zeta_2} \operatorname{sign}\left(\Delta\alpha_p\right).$$

$$(5.92)$$

The next table, Table 5.3, summarizes the nomenclature used for the swash-plate mechanism, where an index A or B always refers to the corresponding hydraulic actuator A or B, respectively.

Table 5.3. Nomenclature for the swash-plate mechanism.

$A_o\left(\Delta\alpha_p\right)$	:	orifice area
A_A, A_B, A	:	effective piston areas
$C_{ext,A}, C_{ext,B}$	:	external laminar leakage coefficients
F_A, F_B	:	actuator forces
F_S, F_{pre}, c_S	:	spring force, prestress-force, spring coefficient
$F_{S,A}, F_{S,B}$	:	forces of free-body-system
$M_{S,fric}, M_S$	:	friction torque, swivel torque
$m_{S,A}, m_{S,B}$	:	masses of piston and piston rod
$p_A, p_B, \Delta p$	:	pressures in the actuator chambers, load pressure
p_T, p_S	:	tank pressure, supply pressure
q_A, q_B	:	flows from and to the chambers
$q_{ext,A}, q_{ext,B}$	:	external laminar leakage flows
$R_{S,eff}$	:	effective radius from the swash-plate pivot to the forces
T, T_0	:	oil temperature, reference temperature
$V_{0,A}, V_{0,B}, V_0$	:	chamber volumes for $\alpha_p = 0$
$\alpha_p, \alpha_{p,d}$	:	swash-plate angle, desired swash-plate angle
$\pm\alpha_{p,\max}$	:	maximum/minimum swash-plate angle (end stops)
$\Delta\alpha_p$	:	swash-plate angle error
$\beta_T\left(T\right)$	:	bulk modulus of oil at temperature T
λ_3	:	characteristic oil coefficient
Θ_S	:	moment of inertia of swash-plate
$\mu\left(T\right)$	:	dynamic viscosity of oil at temperature T
ν_p	:	angular velocity of swash-plate
$\rho_{oil}\left(T\right), \rho_{oil,0}$	:	density of oil at temperature T, T_0

5.5.3 Discrete Open-loop Observer and Measurement Results

The continuous mathematical model (5.92) serves as a basis for a discrete open-loop observer for the swash-plate angle α_p. Let us assume that the quantities supply pressure p_S, temperature T and orifice area $A_o(\Delta\alpha_p)$ are constant during the sampling intervals with the sampling time T_a, i.e., $p_S = p_{S,k}$, $T = T_k$ and $A_o(\Delta\alpha_p) = A_o(\Delta\alpha_{p,k})$ for $kT_a \leq t < (k+1)T_a$, $k = 0, 1, 2, \dots$. Then it is possible to calculate the corresponding exact discrete model to (5.92) in the form

$$\alpha_{p,k+1} = \exp\left(\zeta_{1,k}T_a\right)\alpha_{p,k} + \frac{\zeta_{2,k}\left(\exp\left(\zeta_{1,k}T_a\right) - 1\right)}{\zeta_{1,k}}\operatorname{sign}\left(\Delta\alpha_{p,k}\right) .$$

$$(5.93)$$

The open-loop observer (5.93) was implemented in a drive box for vehicular drive systems developed by STEYR Antriebstechnik GmbH &CO OHG [71]. Fig. 5.14 presents the comparative results of the measured and the estimated swash-plate angle due to (5.93) for a calibration test and for a rapid swash-plate turn with an initial error of 5 ° in the swash-plate angle for a supply pressure $p_S = 28 \cdot 10^5$ Nm^{-2}, an average temperature $T = 60$ °C and a sampling time $T_a = 10 \cdot 10^{-3}$ s. The orifice area as a function of the swash-plate angle error $A_o(\Delta\alpha_p)$ was made available by the manufacturer of the swash-plate mechanism.

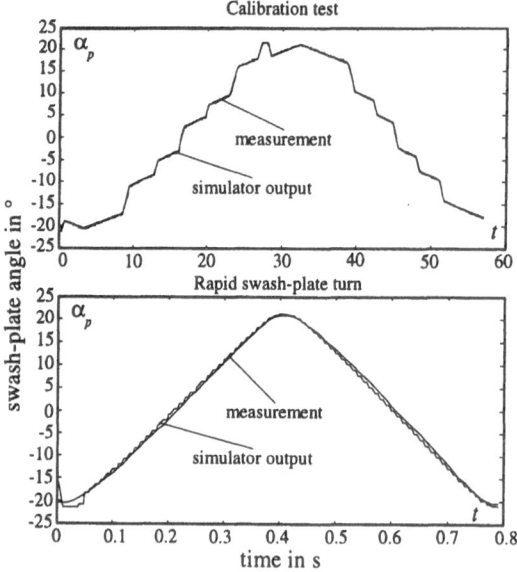

Fig. 5.14. Comparison of the measured and the observed swash-plate angle for a calibration process and a rapid swash-plate turn.

References

1. Abraham R., Marsden J.E., Ratiu T. (1988) *Manifolds, Tensor Analysis and Applications*, 2nd edn., Springer, New York.
2. Aigner M. (1997) *Regelung eines getakteten Schaltnetzteils mittels digitalem Signalprozessor*, MA Thesis, Department of Automatic Control and Control Systems Technology, Johannes Kepler University, Linz, Austria.
3. Aistleitner K., Kugi A., Manhartsgruber B. (1998) On-line identification of deformation resistance and mill stretch coefficient for cold rolling mills, In: *Proc. of METAL'98*, Ostrava, Czech Republic, May 12-14, 1998, Vol.2, 109-116.
4. Anderson G.L., Crowson A., Chandra J. (1992) Introduction to Smart Structures, In: Tzou H.S., Anderson G.L. (Eds.), *Intelligent Structural Systems*, Kluwer Academic Publishers, Dordrecht, 1-8.
5. Athans M., Falb P.L. (1966) *Optimal Control*, McGraw-Hill, New York.
6. Bailey T., Hubbard J.E. (1985) Distributed Piezoelectric-Polymer Active Vibration Control of a Cantilever Beam, *Journal of Guidance and Control*, Vol.8, No.5, 605-611.
7. Balas M.J. (1978) Feedback Control of Flexible Systems, *IEEE Trans. on Automatic Control*, Vol.23, No.4, 673-679.
8. Ball A., Helton W., Walker L. (1993) H_∞-Control for Nonlinear Systems with Output Feedback, *IEEE Trans. on Automatic Control*, Vol.38, No.4, 546-559.
9. Banks H.T., Smith R.C., Wang Y. (1996) *Smart Material Structures: Modeling, Estimation and Control*, John Wiley & Sons, Masson, Paris.
10. Başar T., Bernhard P. (1991) H^∞-*Optimal Control and Related Minmax Design Problems*, Birkhäuser, Boston.
11. Bergen A.R. (1986) *Power Systems Analysis*, Prentice-Hall, New Jersey.
12. Bindel R., Nitsche R., Rothfuß R., Zeitz M. (1999) Flatness based control of a two valve hydraulical joint actuator of a large manipulator, In: *Proc.-CD of the 1999 European Control Conference ECC'99*, Karlsruhe, Germany, August 31 - September 3, 1999, F1009-2.
13. Blackburn J.F., Reethof G., Shearer J.L. (1960) *Fluid Power Control*, John Wiley & Sons, New York and London.
14. Bland D.R., Ford H. (1952) Cold Rolling with Strip Tension, Part III: An Approximate Treatment of the Elastic Compression of the Strip in Rolling Mills, *Journal of the Iron and Steel Institute*, 245-249.
15. Bobrow J.E., Lum K. (1996) Adaptive, High Bandwidth Control of a Hydraulic Actuator, *Journal of Dynamic Systems, Measurement, and Control*, Trans. of the ASME, Vol.118, 714-720.
16. Burke W.L. (1994) *Applied Differential Geometry*, Cambridge University Press, Cambridge.
17. Byrnes C.I., Isidori A., Willems J.C. (1991) Passivity, Feedback Equivalence, and the Global Stabilization of Minimum Phase Nonlinear Systems, *IEEE Trans. on Automatic Control*, Vol.36, No.11, 1228-1240.

18. Choquet-Bruhat Y., D$_E$ Witt-Morette C. (1991), *Analysis, Manifolds and Physics, Part I: Basics*, Revised Edition, Elsevier (North-Holland), Amsterdam.

19. Chua L.O., Desoer C.A., Kuh E.S. (1987) *Linear and Nonlinear Circuits*, McGraw-Hill, New York.

20. Chui C.K., Chen G. (1989), *Linear Systems and Optimal Control*, Springer, Heidelberg.

21. Christofides N. (1975), *Graph Theory: An Algorithmic Approach*, Academic Press, London.

22. Crouch P.E., van der Schaft A.J. (1987) *Variational and Hamiltonian Control Systems*, Lecture Notes in Control and Information Sciences 101, Springer, Berlin, Heidelberg.

23. de Wit C.C., Olsson H., Åström K.J., Lischinsky P. (1995) A New Model for Control of Systems with Friction, *IEEE Trans. on Automatic Control*, Vol.40, No.3, 419-425.

24. Dutton K., Groves C.N. (1996) Self-tuning control of a cold mill automatic gauge control system, *Int. J. Control*, Vol.65, No.4, 573-588.

25. Elasser A., Torrey D.A. (1995) Soft Switching Active Snubbers for DC/DC Converters, *IEEE Trans. on Power Electronics*, Vol.11, No.5, 710-722.

26. Escobar G., van der Schaft A.J., Ortega R. (1999) A Hamiltonian Viewpoint in the Modeling of Switching Power Converters, *Automatica*, Special Issue on Hybrid Systems, Pergamon, Vol.35, 445-452.

27. FitzSimons P.M., Palazzolo J.J. (1996) Part I, II: Modeling of a One-Degree-of-Freedom Active Hydraulic Mount, *Journal of Dynamic Systems, Measurement, and Control*, Trans. of the ASME, Vol.118, 439-448.

28. Fliess M., Glad S.T. (1993) An Algebraic Approach to Linear and Nonlinear Control, In: Trentelman H.L., Willems J.C. (Eds.), *Essays on Control: Perspectives in the Theory and its Applications*, Birkhäuser, Boston, 223-267.

29. Fleck N.A., Johnson K.L., Mear M.E., Zhang L.C. (1992) Cold rolling of foil, *Proc. Instn. Mech. Engrs., Part B: Journal of Engineering Manufacture*, Vol.206, 119-131.

30. Ford H., Ellis F., Bland D.R. (1951) Cold Rolling with Strip Tension, Part I: A New Approximate Method of Calculation and a Comparison with Other Methods, *Journal of the Iron and Steel Institute*, 57-72.

31. Ford H., Ellis F. (1952) Cold Rolling with Strip Tension, Part II: Comparison of Calculated and Experimental Results, *Journal of the Iron and Steel Institute*, 239-245.

32. Frankel T. (1997) *The Geometry of Physics*, Cambridge University Press, New York.

33. Garimella S.S., Srinivasan K. (1996) Application of Repetitive Control to Eccentricity Compensation in Rolling, *Journal of Dynamic Systems, Measurement, and Control*, Trans. of the ASME, Vol.118, 657-664.

34. Ge P., Jouaneh M. (1996) Tracking Control of a Piezoceramic Actuator, *IEEE Trans. on Control Systems Technology*, Vol.4, No.3, 209-215.

35. Ginzburg V.B. (1998), *Roll Eccentricity*, Rolling Mill Technology Series, Volume 1, International Rolling Mill Consultants, Pittsburgh.

36. Goldstein H. (1991), *Klassische Mechanik*, 11.Auflage, AULA-Verlag, Wiesbaden.

37. Greenwood A. (1991), *Electrical Transients in Power Systems*, 2nd edn., John Wiley & Sons, New York.

38. Grimble M.J., Hearns G. (1998) LQG Controllers for State-Space Systems with Pure Transport Delays: Application to Hot Strip Mills, *Automatica*, Vol.34, No.10, 1169-1184.

39. Grimble M.J., Hearns G. (1999) Advanced Control for Hot Rolling Mills, In: Frank P.M. (Ed.), *Advances in Control*, Highlights of ECC'99, Springer, London, 135-169.
40. Haas W., Kugi A., Schlacher K., Paster M. (1998) Experimental Results of the Control of Structures with Piezoelectric Actuators and Sensors, In: *Proc. of the 4th International Conference on Motion and Vibration Control MOVIC'98*, Zürich, Switzerland, August 25-28, 1998, Vol.2, 393-398.
41. Hensel A., Spittel T. (1990) *Kraft- und Arbeitsbedarf bildsamer Formgebungsverfahren*, Deutscher Verlag für Grundstoffindustrie, Leipzig.
42. Hill D., Moylan P. (1976) The Stability of Nonlinear Dissipative Systems, *IEEE Trans. on Automatic Control*, Vol.21, 708-711.
43. Hillerström G., Sternby J. (1996) Adaptive Rejection of Periodic Disturbances with Unknown Period, *Journal of Dynamic Systems, Measurement, and Control*, Trans. of the ASME, Vol.118, 606-610.
44. Hirsch M., Smale S. (1974) *Differential Equations, Dynamical Systems, and Linear Algebra*, Academic Press, San Diego.
45. Höglinger M. (1997) *Aufbau und Simulation eines getakteten Schaltnetzteiles mit Anbindung an einen digitalen Signalprozessor*, MA Thesis, Department of Automatic Control and Control Systems Technology, Johannes Kepler University, Linz, Austria.
46. Hu J., Tomizuka M. (1993) A New Plug-In Adaptive Controller for Rejection of Periodic Disturbances, *Journal of Dynamic Systems, Measurement, and Control*, Trans. of the ASME, Vol.115, 543-546.
47. Imura J., Maeda H., Sugie T., Yoshikawa T. (1995) *Robust stabilization of nonlinear systems by H_∞ state feedback*, Systems & Control Letters 24, Elsevier, 103-114.
48. Incropera F.P., DeWitt D.P. (1996) *Fundamentals of Heat and Mass Transfer*, 4th edn., John Wiley & Sons, New York.
49. Irschik H., Belyaev A.K., Krommer M., Schlacher K. (1997) Non-uniqueness of two inverse problems of thermally and force-loaded smart structures: Sensor shaping and actuator shaping problem, In: *Proc. of the 1997 ASME International Mechanical Engineering Congress and Exposition*, Dallas, USA, November 16-21, 1997, AD-Vol.55, 119-125.
50. Irschik H., Hagenauer K., Ziegler F. (1997) An Exact Solution for Structural Shape Control by Piezoelectric Actuation, In: Gabbert U. (Ed.), *Smart Mechanical Systems-Adaptronics*, Fortschrittberichte VDI, Reihe 11, Nr. 244, VDI-Verlag, Düsseldorf, 93-98.
51. Irschik H., Adam Ch., Heuer R., Ziegler F. (1998) An Exact Solution for Static Shape Control Using Piezoelectric Actuation, In: Bahei-El-Din Y.A., Dvorak G.J. (Eds.), *Proc. of the IUTAM Symposium on Transformation Problems in Composite and Active Materials*, Cairo, Egypt, March 9-12, 1997, Kluwer Academic Publishers, Dordrecht, 247-258.
52. Isidori A. (1996) *Nonlinear Control Systems*, 3rd edn., Springer, London.
53. Isidori A., Astolfi A. (1992) Disturbance Attenuation and H_∞-Control via Measurement Feedback in Nonlinear Systems, *IEEE Trans. on Automatic Control*, Vol.37, No.9, 1283-1293.
54. Ivantysyn J., Ivantysyn M. (1993) *Hydrostatische Pumpen und Motoren*, Vogel Buchverlag, Würzburg.
55. Joshi S.M. (1988), *Control of Large Flexible Structures*, Lecture Notes in Control and Information Sciences 131, Springer, London.
56. Kallenbach E., Eick R., Quendt P. (1994) *Elektromagnete*, B.G. Teubner Verlag, Stuttgart.

57. Kassakian J.G., Schlecht M.F., Verghese G.C. (1992) *Principles of Power Electronics*, Addison Wesley, New York.
58. Knobloch H.W., Isidori A., Flockerzi D. (1993) *Topics in Control Theory*, DMV Seminar Band 22, Birkhäuser, Basel.
59. Khalil H.K. (1992) *Nonlinear Systems*, Macmillan Publishing Company, New York.
60. Krstić M., Kanellakopoulos I., Kokotović P. (1995) *Nonlinear and Adaptive Control Design*, John Wiley & Sons, New York.
61. Kugi A., Schlacher K., Kaltenbacher M. (1996) Object Oriented Approach for Large Circuits with Substructures in the Computer Algebra Program Maple V, In: Silvester P.P. (Ed.), *Software for Electrical Engineering Analysis and Design*, Computational Mechanics Publications, Southampton, 491-500.
62. Kugi A., Schlacher K., Irschik H. (1997) Optimal Control of Nonlinear Parametrically Excited Beam Vibrations By Spatially Distributed Sensors and Actuators, In: *Proc.-CD of the 1997 ASME Design Engineering Technical Conferences DETC'97*, Sacramento, USA, September 14-17, 1997, VIB-4171.
63. Kugi A., Schlacher K., Angerer R., Helekal G. (1997) Nonlinear Temperature Control of Sheet Metal Specimen, In: *Proc. of the Workshop Automation 2001 (Tempus JEP 07759)*, Vienna, Austria, June 12-14, 181-190.
64. Kugi A., Haas W., Aistleitner K. (1998) Method for Compensating Eccentricity of Supporting and/or Working Roller in Dual or Quadruple Roll Housing, *International patent pending*, June 11, 1998, WO 98/24567.
65. Kugi A., Frank H., Aistleitner K. (1999) Method and Device for the Active Compensation of Periodic Disturbances During Hot or Cold Rolling, *European patent pending*, October 7, 1999, EP 0 992 295 A2.
66. Kugi A., Schlacher K. (1999) Nonlinear H_∞-Controller Design for a DC-to-DC Power Converter, *IEEE Trans. on Control Systems Technology*, Vol.7, No.2, 230-237.
67. Kugi A., Schlacher K., Keintzel G. (1999) Position Control and Active Eccentricity Compensation in Rolling Mills, *at Automatisierungstechnik* 47, Oldenbourg, 8/99, 342-349.
68. Kugi A., Schlacher K., Irschik H. (1999) Infinite Dimensional Control of Nonlinear Beam Vibrations by Piezoelectric Actuator and Sensor Layers, *Nonlinear Dynamics*, Kluwer, Vol.19, No.1, 71-91.
69. Kugi A., Schlacher K., Kaltenbacher M., Lerch R. (1999) Energy Shaping Control of Electrostatic Membrane Vibrations, In: *Proc.-CD of DETC'99, ASME Design Engineering Technical Conferences*, Las Vegas, USA, September 12-15, 1999, VIB-8010.
70. Kugi A., Schlacher K., Novak R. (1999) Software Package: Non-linear Affine-Input Control Systems, *Maple-Application Center, Engineering, Control*, http://www.maplesoft.com.
71. Kugi A., Schlacher K., Aitzetmüller H., Hirmann G. (2000) Modeling and Simulation of a Hydrostatic Transmission with Variable-Displacement Pump, In: *Proc. of the 3rd Mathmod Vienna*, Vienna, Austria, February 2-4, 2000, Vol.1, 99-102.
72. Kugi A. (2000) Regelung eines Ćuk-Konverters, *at Automatisierungstechnik* 48, Oldenbourg, 3/2000, 116-123.
73. Kugi A., Haas W., Schlacher K., Aistleitner A., Frank H., Rigler G.W. (2000) Active Compensation of Roll Eccentricity in Rolling Mills, *IEEE Trans. on Industry Applications*, Vol.36, No.2, 625-632.
74. Kugi A., Novak R., Schlacher K., Aistleitner K. (2000) A Flatness Based Approach for the Thickness Control in Rolling Mills, In: *Proc.-CD of the Fourteenth*

International Symposium of Mathematical Theory of Networks and Systems, Perpignan, France, June 19-23, 2000.

75. Lee C.-K., Moon F.C. (1990) Modal Sensors/Actuators, *Journal of Applied Mechanics,* Trans. of the ASME, Vol.57, 434-441.

76. Lee C.-K. (1992) Piezoelectric Laminates: Theory and Experiments for Distributed Sensors and Actuators, In: Tzou H.S., Anderson G.L. (Eds.), *Intelligent Structural Systems,* Kluwer Academic Publishers, Dordrecht, 75-167.

77. Lemmen M., Bröcker M. (1999) Nonlinear Control of Hydraulic Diffrential Cylinders, In: *Preprints of the first Nonlinear Control Network (NCN),* Pedagogical School, Athens, Greece, September 6-10, 1999, 441-444.

78. Lennevi J. (1995) Hydrostatic Transmission Control, Design Methodology for Vehicular Drivetrain Applications, *Linköping Studies in Science and Technology,* Dissertations, No.395, Linköping, Sweden.

79. Leung F.H.F, Tam P.K.S., Li C.K. (1993) An Improved LQR-based Controller for Switching DC-dc Converters, *IEEE Trans. on Industrial Electronics,* Vol.40, No.5, 521-528.

80. Lewin J. (1994) *Differential Games,* Springer, London.

81. Lide R.D. (1995) *Handbook of Chemistry and Physics,* 76th edn. 1995-1996, CRC Press, Boca Raton, Florida.

82. Liu Z., Zheng S. (1999) *Semigroups associated with dissipative systems,* Chapman & Hall/CRC Press, Boca Raton, London.

83. Liu X. Z., Verghese G.C., Lang J.H., Önder M.K. (1989) Generalizing the Blondel-Park Transformation of Electrical Machines: Necessary and Sufficient Conditions, *IEEE Trans. on Circuits and Systems,* Vol.36, No.8, 1058-1066.

84. Lu W.-M., Packard A. (1996) Adaptive H_∞-Control for Nonlinear Systems: A Dissipation Theoretical Approach, *CDS Technical Reports,* CIT/CDS 96-019, www.cds.caltech.edu/reports.

85. MacFarlane A.G.J. (1970) Formulation of the state-space equations for nonlinear networks, *Int. Journal of Control,* Vol.11, No.3, 449-470.

86. Manhartsgruber B. (1999) Singular Pertubation Analysis of an Electrohydraulic Servo-Drive with Discontinuous Reduced Dynamics, In: *Proc.-CD of the 1999 ASME Design Engineering Technical Conferences DETC'99,* Las Vegas, USA, September 12-16, 1999, DETC99/VIB-8178.

87. Manring N.D., Johnson R.E. (1996) Modeling and Designing a Variable-Displacement Open-Loop Pump, *Journal of Dynamic Systems, Measurement, and Control,* Trans. of the ASME, Vol.118, 267-271.

88. Manring N.D. (1997) The Effective Fluid Bulk-Modulus Within a Hydrostatic Transmission, *Journal of Dynamic Systems, Measurement, and Control,* Trans. of the ASME, Vol.119, 462-466.

89. Manring N.D., Luecke G.R. (1998) Modelling and Designing a Hydrostatic Transmission with a Fixed-Displacement Motor, *Journal of Dynamic Systems, Measurement, and Control,* Trans. of the ASME, Vol.120, 45-49.

90. Manring N.D. (1998) The Torque on the Input Shaft of an Axial-Piston Swash-Plate Type Hydrostatic Pump, *Journal of Dynamic Systems, Measurement, and Control,* Trans. of the ASME, Vol.120, 57-62.

91. Marino R., Tomei P. (1995) *Nonlinear Control Design,* Prentice Hall, London.

92. Marsden J.E., Hughes T.J.R. (1994) *Mathematical Foundations of Elasticity,* Dover Publications, New York.

93. Marsden J.E., Ratiu T.S. (1999) *Introduction to Mechanics and Symmetry,* 2nd edn., Springer, New York.

94. Massimo F.M., Kwatny H.G., Bahar L.Y. (1980) Derivation of Brayton-Moser Equations from a Topological Mixed Potential Function, *Journal of The Franklin Institute,* Pergamon Press, Vol.310, No.4/5, 259-269.

95. Meirovitch L., Baruh H. (1985) The Implementation of Modal Filters for Control of Structures, *Journal of Guidance and Control*, Vol.8, No.6, 707-716.
96. Merritt H.E. (1967) *Hydraulic Control Systems*, John Wiley & Sons, New York.
97. Middlebrook R.D., Ćuk S. (1976) A general unified approach to modeling switching-converter power stages, In: *Proc. of the IEEE Power Electronics Specialists Conference, IEEE Publication 76 CHO 1084-3AES*, 18-34.
98. Miller S.E., Oshman Y., Abramovich H. (1996) Modal Control of Piezolaminated Anisotropic Rectangular Plates Part 1: Modal Transducer Theory, Part 2: Control Theory, *AIAA Journal*, Vol.34, No.9, 1868-1875, 1876-1884.
99. Mohan N., Undeland T.M., Robbins W.P. (1989) *Power Electronics: Converters, Applications, and Design*, John Wiley & Sons, New York.
100. Moylan J.P. (1974) Implications of Passivity in a Class of Nonlinear Systems, *IEEE Trans. on Automatic Control*, Vol.AC-19, No.4, 373-381.
101. Murrenhoff H. (1997) *Grundlagen der Fluidtechnik, Teil 1: Hydraulik*, Wissenschaftsverlag, Aachen.
102. Nakagawa S., Miura H., Fukushima S., Amasaki J. (1990) Gauge Control System For Hot Strip Finishing Mill, In: *Proc. of the 29th Conference on Decision and Control*, Honolulu, Hawaii, December 1990, TA-8-2-10:40, 1573-1578.
103. Nijmeijer H., van der Schaft A.J. (1991) *Nonlinear Dynamical Control Systems*, Springer, New York.
104. Novak R., Schlacher K., Kugi A., Frank H. (2000) Nonlinear Hydraulic Gap Control: A Practical Approach, In: *Preprints of the IFAC Conference on Control Systems Design (CSD2000)*, Bratislava, Slovak Republic, June 18-20, 2000, 605-609.
105. Nowacki W. (1975) *Dynamic Problems of Thermoelasticity*, Noordhoff International Publishing, PWN-Polish Scientific Publishers, Warszawa.
106. Olver P.J., (1993) *Applications of Lie Groups to Differential Equations*, Springer, New York.
107. Ortega R., Loria A., Nicklasson P.J., Sira Ramírez H. (1998) *Passivity-based Control of Euler-Lagrange Systems*, Springer, London.
108. Ortega R., van der Schaft A.J., Maschke B., Escobar G. (2000) Interconnection and Damping Assignment Passivity-based Control of Port-Controlled Hamiltonian Systems, *Automatica*, Pergamon, in press.
109. Ortega R., van der Schaft A.J., Mareels I., Maschke B. (2000) Putting Energy Back in Control, *Automatica*, Pergamon, in press.
110. Pedersen L.M. (1999) *Modeling and Control of Plate Mill Processes*, Published by Department of Automatic Control, Lund Institute of Technology, Box 118, SE-221 00 Lund, Sweden.
111. Plummer A.R., Vaughan N.D. (1996) Robust Adaptive Control for Hydraulic Servosystems, *Journal of Dynamic Systems, Measurement, and Control*, Trans. of the ASME, Vol.118, 237-244.
112. Rigler G.W., Aberl H.R., Staufer W., Aistleitner K., Weinberger K.H. (1996) Improved Rolling Mill Automation by Means of Advanced Control Techniques and Dynamic Simulation, *IEEE Trans. on Industry Applications*, Vol.32, No.3, 599-607.
113. Sabonović A., Sabonović N., Ohnishi K. (1993) Sliding modes in power converters and motion control systems, *Int. Journal of Control*, Vol.57, No.5, 1237-1259.
114. Salama M.H., Holmes P.G. (1996) Transient and steady-state load performance of a stand-alone self-excited induction generator, *IEE Proc. Electr. Power Appl.*, Vol.143, No.1, 50-58.
115. Sastry S. (1999) *Nonlinear Systems, Analysis, Stability, and Control*, Springer, New York.

116. Schlacher K., Kugi A. (1994) Modern Control of a Čuk-Converter Using Non-linear Methods, In: *Proc. of the 3rd IEEE Conference on Control Applications*, Glasgow, UK, August 24-26, 1994, Vol.1, 503-504.
117. Schlacher K., Irschik H., Kugi A. (1997) Control of Nonlinear Beam Vibrations by Multiple Piezoelectric Layers, In: van Campen D.H. (Ed.), *Proc. of the IUTAM Symposium on Interaction between Dynamics and Control in Advanced Mechanical Systems*, Eindhoven, The Netherlands, April 21-26, 1996, Kluwer Academic Publishers, Dordrecht, 355-362.
118. Schlacher K., Kugi A., Irschik H. (1996) H_∞-Control of Nonlinear Beam Vibrations, In: *Proc. of the 3rd International Conference on Motion and Vibration Control MOVIC'96*, Chiba-Tokyo, Japan, September 1-6, 1996, Vol.3, 497-484.
119. Schlacher K., Kugi A. (1997) Regelung von Hamiltonsystemen, *e&i* 114, H.7/8, Springer, 353-359.
120. Schlacher K., Kugi A., Scheidl R. (1998) Tensor analysis based symbolic computation for mechatronic systems, *Mathematics and Computers in Simulation* 46, Elsevier, 517-525.
121. Schlacher K., Kugi A., Haas W. (1998) Geometric Control of a Class of Nonlinear Descriptor Systems, In: Huijberts J., Nijmeijer H., van der Schaft A., Scherpen J. (Eds.), *Preprints of the 4th IFAC Nonlinear Control Systems Design NOLCOS '98*, Enschede, The Netherlands, July 1-3, 1998, Vol.2, 387-392.
122. Schlacher K., Kugi A. (1999) Stabilization of Mechanical Structures with Distributed Sensors and Actuators: A Hamiltonian Approach, In: *Proc.-CD of the 1999 ASME Design Engineering Technical Conferences DETC'99*, Las Vegas, USA, September 12-16, 1999, DETC99/VIB-8008.
123. Schlacher K., Kugi A. (1999) Control of Mechanical Strucutures by Piezoelectric Actuators and Sensors, In: Aeyels D., Lamnabhi-Lagarrique F., van der Schaft A. (Eds.) *Stability and Stabilization of Nonlinear Systems*, Lecture Notes in Control and Information Sciences 246, Springer, London, 275-291.
124. Schlacher K., Haas W., Kugi A. (1999) Ein Vorschlag für eine Normalform von Deskriptorsystemen, *ZAMM Zeitschrift für angewandte Mathematik und Mechanik*, Vol.79, John Wiley, 21-24.
125. Schlacher K., Kugi A. (2000) Control of Elastic Systems, A Hamiltonian Approach, In: Leonard N.E., Ortega R. (Eds.) *Preprints of the Workshop on Lagrangian and Hamiltonian Methods for Nonlinear Control*, Princeton, New Jersey USA, March 16-18, 2000, 80-85.
126. Sepulchre R., Janković M., Kokotović P. (1997) *Constructive Nonlinear Control*, Springer, London.
127. Sira Ramírez H. (1989) A Geometric Approach to Pulse-Width Modulated Control in Nonlinear Dynamical Systems, *IEEE Trans. on Automatic Control*, Vol.34, No.2, 184-187.
128. Sira Ramírez H. (1991) Nonlinear P-I Controller Design for Switchmode dc-to-dc Power Converters, *IEEE Trans. on Circuits and Systems*, Vol.38, No.2, 410-417.
129. Sira Ramírez H., Lischinsky-Arenas P. (1991) Differential algebraic approach in non-linear dynamical compensator design for d.c.-to-d.c. power converters, *Int. Journal of Control*, Vol.54, No.1, 111-133.
130. Sira Ramírez H., Prada-Rizzo M.T. (1992) Nonlinear Feedback Regulator Design for the Čuk Converter, *IEEE Trans. on Automatic Control*, Vol.37, No.8, 1173-1180.
131. Sira Ramírez H., Tarantino-Alvarado R., Llanes-Santiago O. (1993) Adaptive feedback stabilization in PWM-controlled DC-to-DC power supplies, *Int. Journal of Control*, Vol.57, No.3, 599-625.

132. Sira Ramírez H., Perez-Moreno R.A., Ortega R., Garcia-Esteban M. (1997) Passivity-Based Controllers for Stabilization of Dc-to-DC Power Converters, *Automatica*, Pergamon, Vol.33, No.4, 499-513.

133. Sira Ramírez H. (1998) A general canonical form for feedback passivity of nonlinear systems, *Int. Journal of Control*, Vol.71, No.5, 891-905.

134. Sievers L.A., von Flotow A.H. (1992) Comparison and Extensions of Control Methods for Narrow-Band Disturbance Rejection, *IEEE Trans. on Signal Processing*, Vol.40, No.10, 2377-2391.

135. Slotine J.-J. E., Li W. (1991) *Applied Nonlinear Control*, Prentice Hall, New Jersey.

136. Tafazoli S., de Silva C.W., Lawrence P.D. (1998) Tracking Control of an Electrohydraulic Manipulator in the Presence of Friction, *IEEE Trans. on Control Systems Technology*, Vol.6, No.3, 401-411.

137. Tauchert T.R. (1992) Piezothermoelastic Behavior of a Laminated Plate, *Journal of Thermal Stresses* 15, 25-37.

138. Thoma J. (1970) Mathematische Modelle und die effektive Leistung hydrostatischer Maschinen und Getriebe, *Ölhydraulik und Pneumatik* 14, Nr.6, 233-237.

139. Tzou H.S., Anderson G.L. (1992) *Intelligent Structural Systems*, Kluwer Academic Publishers, Dordrecht.

140. Tzou H.S., Zhong J.P., Hollkamp J.J. (1994) Spatially Distributed Orthogonal Piezoelectric Shell Actuators: Theory and Applications, *Journal of Sound and Vibration*, Vol.177, No.3, 363-378.

141. van der Schaft A.J. (1992) L_2-Gain Analysis of Nonlinear Systems and Nonlinear State Feedback H_∞ Control, *IEEE Trans. on Automatic Control*, Vol.37, No.6, 770-784.

142. van der Schaft A.J. (1993) Nonlinear State Space H_∞ Control Theory, In: Trentelman H.L., Willems J.C. (Eds.), *Essays on Control: Perspectives in the Theory and its Applications*, Birkhäuser, Boston, 153-190.

143. van der Schaft A.J. (2000) L_2-Gain and Passivity Techniques in Nonlinear Control, Springer, London.

144. Vidyasagar M. (1993) *Nonlinear Systems Analysis*, 2nd edn., Prentice Hall, New Jersey.

145. Willems J.C. (1972) Dissipative dynamical systems, Part I: General Theory, *Arch. Rational Mech. Anal.*, Vol.45, 321-351.

146. Wilson W.E., Lemme C.D. (1970) Hydrostatic Transmissions, Part 2 - Ideal and Real Performance, *Hydraulics & Pneumatics* May 1970, 90-95.

147. Ziegler F. (1995) *Mechanics of Solids and Fluids*, 2nd edn., Springer, New York.

Index

Lecture Notes in Control and Information Sciences

Edited by M. Thoma and M. Morari

1997–2000 Published Titles:

Vol. 222: Morse, A.S.
Control Using Logic-Based Switching
288 pp. 1997 [3-540-76097-0]

Vol. 223: Khatib, O.; Salisbury, J.K.
Experimental Robotics IV: The 4th
International Symposium, Stanford,
California,
June 30 - July 2, 1995
596 pp. 1997 [3-540-76133-0]

Vol. 224: Magni, J.-F.; Bennani, S.;
Terlouw, J. (Eds)
Robust Flight Control: A Design Challenge
664 pp. 1997 [3-540-76151-9]

Vol. 225: Poznyak, A.S.; Najim, K.
Learning Automata and Stochastic
Optimization
219 pp. 1997 [3-540-76154-3]

Vol. 226: Cooperman, G.; Michler, G.;
Vinck, H. (Eds)
Workshop on High Performance Computing
and Gigabit Local Area Networks
248 pp. 1997 [3-540-76169-1]

Vol. 227: Tarbouriech, S.; Garcia, G. (Eds)
Control of Uncertain Systems with Bounded
Inputs
203 pp. 1997 [3-540-76183-7]

Vol. 228: Dugard, L.; Verriest, E.I. (Eds)
Stability and Control of Time-delay Systems
344 pp. 1998 [3-540-76193-4]

Vol. 229: Laumond, J.-P. (Ed.)
Robot Motion Planning and Control
360 pp. 1998 [3-540-76219-1]

Vol. 230: Siciliano, B.; Valavanis, K.P. (Eds)
Control Problems in Robotics and
Automation
328 pp. 1998 [3-540-76220-5]

Vol. 231: Emel'yanov, S.V.; Burovoi, I.A.;
Levada, F.Yu.
Control of Indefinite Nonlinear Dynamic
Systems
196 pp. 1998 [3-540-76245-0]

Vol. 232: Casals, A.; de Almeida, A.T. (Eds)
Experimental Robotics V: The Fifth
International Symposium Barcelona,
Catalonia, June 15-18, 1997
190 pp. 1998 [3-540-76218-3]

Vol. 233: Chiacchio, P.; Chiaverini, S. (Eds)
Complex Robotic Systems
189 pp. 1998 [3-540-76265-5]

Vol. 234: Arena, P.; Fortuna, L.; Muscato, G.;
Xibilia, M.G.
Neural Networks in Multidimensional
Domains: Fundamentals and New Trends in
Modelling and Control
179 pp. 1998 [1-85233-006-6]

Vol. 235: Chen, B.M.
H∞ Control and Its Applications
361 pp. 1998 [1-85233-026-0]

Vol. 236: de Almeida, A.T.; Khatib, O. (Eds)
Autonomous Robotic Systems
283 pp. 1998 [1-85233-036-8]

Vol. 237: Kreigman, D.J.; Hagar, G.D.;
Morse, A.S. (Eds)
The Confluence of Vision and Control
304 pp. 1998 [1-85233-025-2]

Vol. 238: Elia, N. ; Dahleh, M.A.
Computational Methods for Controller Design
200 pp. 1998 [1-85233-075-9]

Vol. 239: Wang, Q.G.; Lee, T.H.; Tan, K.K.
Finite Spectrum Assignment for Time-Delay
Systems
200 pp. 1998 [1-85233-065-1]